Provost Joe B. Whitehead, Jr., Ph.D.
As we celebrate 125 years of excellence, innovation and pride, we present this book to you as an example of collaborative interdisciplinary research by our faculty and students.

Professor Godfrey A. Uzochukwu, Ph.D.
For Faculty and Students

Proceedings of the 2013 National Conference on Advances in Environmental Science and Technology

Godfrey A. Uzochukwu • Keith Schimmel
Vinayak Kabadi • Shoou-Yuh Chang
Tanya Pinder • Salam A. Ibrahim
Editors

Proceedings of the 2013 National Conference on Advances in Environmental Science and Technology

Editors
Godfrey A. Uzochukwu
North Carolina A&T State University
Greensboro, NC, USA

Keith Schimmel
North Carolina A&T State University
Greensboro, NC, USA

Vinayak Kabadi
North Carolina A&T State University
Greensboro, NC, USA

Shoou-Yuh Chang
North Carolina A&T State University
Greensboro, NC, USA

Tanya Pinder
North Carolina A&T State University
Greensboro, NC, USA

Salam A. Ibrahim
North Carolina A&T State University
Greensboro, NC, USA

ISBN 978-3-319-19922-1   ISBN 978-3-319-19923-8   (eBook)
DOI 10.1007/978-3-319-19923-8

Library of Congress Control Number: 2015946565

Springer Cham Heidelberg New York Dordrecht London
© Springer International Publishing Switzerland 2016
This work is subject to copyright. All rights are reserved by the Publisher, whether the whole or part of the material is concerned, specifically the rights of translation, reprinting, reuse of illustrations, recitation, broadcasting, reproduction on microfilms or in any other physical way, and transmission or information storage and retrieval, electronic adaptation, computer software, or by similar or dissimilar methodology now known or hereafter developed.
The use of general descriptive names, registered names, trademarks, service marks, etc. in this publication does not imply, even in the absence of a specific statement, that such names are exempt from the relevant protective laws and regulations and therefore free for general use.
The publisher, the authors and the editors are safe to assume that the advice and information in this book are believed to be true and accurate at the date of publication. Neither the publisher nor the authors or the editors give a warranty, express or implied, with respect to the material contained herein or for any errors or omissions that may have been made.

Printed on acid-free paper

Springer International Publishing AG Switzerland is part of Springer Science+Business Media (www.springer.com)

Preface

This book contains peer-reviewed chapters accepted for presentation at the National Conference on Advances in Environmental Science and Technology. The chapters are arranged by topics with names of authors and affliations.

Several conversations about environmental regulations, groundwater remediation technologies and waste to energy, climate change, economics and environmental justice, fate and transport of contaminants, food bio-processing, innovative environmental technologies, sustainable energy and water resources and waste management among federal agencies, private agencies, and university professors set the stage for the September 12, 2013 National Conference on Advances in Environmental Science and Technology. The purpose of the National Conference on Advances in Environmental Science and Technology which was held in Greensboro, North Carolina, was to provide a forum for agencies to address advances in environmental science and technology including problems, solutions, and research needs. Our goal was to foster relationships that could result in partnerships needed to protect, sustain the environment and improve the quality of life.

The National Conference on Advances in Environmental Science and Technology was sponsored by Sullivan International Group, Waste Industries, CDM Smith, United States Department of Energy, United States Environmental Protection Agency, National Aeronautics and Space Administration, National Science Foundation, and North Carolina Agricultural and Technical State University. These agencies are thanked for their financial and logistics support. The hard work of Tarcy Keyes, Stephen Johnson, Angela Smith, and Pat O'Connor is gratefully acknowledged. Special thanks to Johnsely S. Cyrus, Stephanie Luster Teasley, and Heather Stewart for their assistance. The following keynote speakers are thanked for their contributions: Michael Maloy, Vice President, Sullivan International Group, San Diego; Greg Green, Director of Outreach and Information, United States Environmental Protection Agency; Barry Edwards, Director of Utilities and Engineering, Catawba County Government, NC; Joe B. Whitehead, Jr., Provost and Vice Chancellor for Academic Affairs at North Carolina Agricultural and Technical State University; and Barry Burks, Vice Chancellor for Research and Economic Development at North Carolina Agricultural and Technical State University.

Greensboro, NC

Godfrey A. Uzochukwu
Vinayak Kabadi
Shoou Yuh Chang
Keith Schimmel
Tanya Pinder
Salam A. Ibrahim

Contents

Part I Climate Change

Assessment of Climate Change Impact on Watershed Hydrology 3
Somsubhra Chattopadhyay and Manoj K. Jha

Trend of Climate Variability in North Carolina During the Past Decades 13
Mohammad Sayemuzzaman and Manoj K. Jha

Arctic Storm and Its Impact on the Surface Winds over the Chukchi-Beaufort Seas 21
Wei Tao, Jing Zhang, Xiangdong Zhang, and Junming Chen

Designing a Remote In Situ Soil Moisture Sensor Network for Small Satellite Data Retrieval 35
Rawfin Zaman, William W. Edmonson, and Manoj K. Jha

Alaskan Regional Climate Changes in Dynamically Downscaled CMIP5 Simulations 47
Jing Zhang, Jeremy Krieger, Uma Bhatt, Chuhan Lu, and Xiangdong Zhang

Part II Fate and Transport of Contaminants

Application of Kalman Filter Embedded with Neural Network in 3-Dimensional Subsurface Contaminant Transport Modeling 63
Godwin Appiah Assumaning and Shoou-Yuh Chang

Application of Adaptive Extended Kalman Filtering Scheme to Improve the Efficiency of a Groundwater Contaminant Transport Model 75
Shoou-Yuh Chang and Elvis B. Addai

Application of Ensemble Square Root Kalman Filter in a Three-Dimensional Subsurface Contaminant Transport Model 87
Shoou-Yuh Chang and Torupallab Ghoshal

Application of 3D VAR Kalman Filter in a Three-Dimensional Subsurface Contaminant Transport Model for a Continuous Pollutant Source 97
Shoou-Yuh Chang and Anup Saha

Groundwater Flow Modeling in the Shallow Aquifer of Buffalo Creek, Greensboro 105
Jenberu Feyyisa, Manoj K. Jha, and Shoou-Yuh Chang

Part III   Food Bioprocessing

Inactivation of *E. coli* O157:H7 on Rocket Leaves by Eucalyptus and Wild-Thyme Essential Oils 119
Saddam S. Awaisheh

Decontamination of *Escherichia coli* O157:H7 from Leafy Green Vegetables Using Ascorbic Acid and Copper Alone or in Combination with Organic Acids 131
Rabin Gyawali and Salam A. Ibrahim

Enzymatic Activity of *Lactobacillus* Grown in a Sweet Potato Base Medium 137
Saeed A. Hayek and Salam A. Ibrahim

Effect of Metal Ions on the Enzymatic Activity of *Lactobacillus reuteri* Growing in a Sweet Potato Medium 145
Saeed A. Hayek and Salam A. Ibrahim

Using Sweet Potatoes as a Basic Component to Develop a Medium for the Cultivation of Lactobacilli 157
Saeed A. Hayek, Abolghasem Shahbazi, and Salam A. Ibrahim

Impact of Gums on the Growth of Probiotic Microorganisms 165
Bernice D. Karlton-Senaye and Salam A. Ibrahim

Interaction Between *Bifidobacterium* and Medical Drugs 171
Temitayo O. Obanla, Saeed A. Hayek, Rabin Gyawali, and Salam A. Ibrahim

Functional Food Product Development from Fish Processing By-products Using Isoelectric Solubilization/Precipitation 179
Reza Tahergorabi and Salam Ibrahim

Part IV Sustainable Energy

Optimizing the Design of Chilled Water Plants in Large Commercial Buildings 187
Dante' Freeland, Christopher Hall, and Nabil Nassif

Optimizing Ice Thermal Storage to Reduce Energy Cost 195
Christopher L. Hall, Dante' Freeland, and Nabil Nassif

Optimization of HVAC Systems Using Genetic Algorithm 203
Tony Nguyen and Nabil Nassif

Artificial Intelligent Approaches for Modeling and Optimizing HVAC Systems 211
Raymond Tesiero, Nabil Nassif, and Harmohindar Singh

Part V Waste Management

Experimental Study of MSW Pyrolysis in Fixed Bed Reactor 223
Emmanuel Ansah, John Eshun, Lijun Wang, Abolghasem Shahbazi, and Guidgopuram B. Reddy

Separate Hydrolysis and Fermentation of Untreated and Pretreated Alfalfa Cake to Produce Ethanol 233
Shuangning Xiu, Nana Abayie Boakye-Boaten, and Abolghasem Shahbazi

Tax Policy's Role in Promoting Sustainability 241
Gwendolyn McFadden and Jean Wells

Scrap Tires Air Emissions in North Carolina 249
Vereda Johnson Williams and Godfrey A. Uzochukwu

Index 259

Contributors

Elvis B. Addai Department of Civil Engineering, North Carolina Agricultural &Technical State University, Greensboro, NC, USA

Emmanuel Ansah Department of Natural Resources and Environmental Design, North Carolina Agricultural and Technical State University, Greensboro, NC, USA

Godwin Appiah Assumaning Department of Civil and Environmental Engineering, NC A&T State University, Greensboro, NC, USA

Saddam S. Awaisheh Department of Nutrition and Food Processing, Al-Balqa Applied University, Salt, Jordan

Uma Bhatt Geophysical Institute, University of Alaska Fairbanks, Fairbanks, AK, USA

Nana Abayie Boakye-Boaten Department of Natural Resources and Environmental Design, North Carolina A and T State University, Greensboro, NC, USA

Shoou-Yuh Chang Department of Civil, Architectural, and Environmental Engineering, North Carolina Agricultural and Technical State University, Greensboro, NC, USA

Somsubhra Chattopadhyay Department of Computational Science and Engineering and Civil, Architectural and Environmental Engineering, North Carolina A&T State University, Greensboro, NC, USA

Junming Chen International Arctic Research Center, University of Alaska Fairbanks, Fairbanks, AK, USA

William W. Edmonson Department of Electrical and Computer Engineering, North Carolina A&T State University, Greensboro, NC, USA

John Eshun Department of Natural Resources and Environmental Design, North Carolina Agricultural and Technical State University, Greensboro, NC, USA

Jenberu Feyyisa Department of Civil, Architectural, and Environmental Engineering, North Carolina Agricultural and Technical State University, Greensboro, NC, USA

Dante' Freeland Department of CAAE Engineering, North Carolina A&T State University, Greensboro, NC, USA

Torupallab Ghoshal Department of Civil Engineering, North Carolina A&T State University, Greensboro, NC, USA

Rabin Gyawali Department of Energy and Environmental Systems, North Carolina Agricultural and Technical State University, Greensboro, NC, USA

Christopher L. Hall Department of Civil and Architectural Engineering, North Carolina A&T State University, Greensboro, NC, USA

Saeed A. Hayek Department of Energy and Environmental Systems, North Carolina Agricultural and Technical State University, Greensboro, NC, USA

Regine Hock Geophysical Institute, University of Alaska Fairbanks, Fairbanks, AK, USA

Salam A. Ibrahim Department of Family and Consumer Sciences, North Carolina Agricultural and Technical State University, Greensboro, NC, USA

Manoj K. Jha Department of Civil, Architectural and Environmental Engineering, North Carolina A&T State University, Greensboro, NC, USA

Bernice D. Karlton-Senaye North Carolina Agricultural and Technical State University, Greensboro, NC, USA

Jeremy Krieger Arctic Region Supercomputing Center, University of Alaska Fairbanks, Fairbanks, AK, USA

Chuhan Lu International Arctic Research Center, University of Alaska Fairbanks, Fairbanks, AK, USA

Gwendolyn McFadden North Carolina A&T State University, Greensboro, NC, USA

Nabil Nassif Department of Civil and Architectural Engineering, North Carolina A&T State University, Greensboro, NC, USA

Tony Nguyen Department of Civil, Architectural and Environmental Engineering, North Carolina A&T State University, Greensboro, NC, USA

Temitayo O. Obanla North Carolina Agricultural and Technical State University, Greensboro, NC, USA

Guidgopuram B. Reddy Department of Natural Resources and Environmental Design, North Carolina Agricultural and Technical State University, Greensboro, NC, USA

Anup Saha Department of Civil Engineering, North Carolina A&T State University, Greensboro, NC, USA

Mohammad Sayemuzzaman Department of Energy and Environmental System, North Carolina A&T State University, Greensboro, NC, USA

Abolghasem Shahbazi Department of Natural Resources and Environmental Design, North Carolina Agricultural and Technical State University, Greensboro, NC, USA

Harmohindar Singh Department of Civil and Architectural Engineering, North Carolina A&T State University, Greensboro, NC, USA

Reza Tahergorabi North Carolina Agricultural and Technical State University, Greensboro, NC, USA

Wei Tao Department of Physics and EES, North Carolina A&T State University, Greensboro, NC, USA

Raymond Tesiero Department of Computational Science and Engineering, North Carolina A&T State University, Greensboro, NC, USA

Godfrey A. Uzochukwu Waste Management Institute, North Carolina A&T State University, Greensboro, NC, USA

Lijun Wang Department of Natural Resources and Environmental Design, North Carolina Agricultural and Technical State University, Greensboro, NC, USA

Jean Wells Howard University, Washington, DC, USA

Vereda Johnson Williams North Carolina A&T State University, Greensboro, NC, USA

Shuangning Xiu Department of Natural Resources and Environmental Design, North Carolina A and T State University, Greensboro, NC, USA

Rawfin Zaman Department of Electrical and Computer Engineering, North Carolina A&T State University, Greensboro, NC, USA

Jing Zhang Department of Physics, North Carolina A&T State University, Greensboro, NC, USA

Department of Energy and Environmental Systems, North Carolina A&T State University, Greensboro, NC, USA

Xiangdong Zhang International Arctic Research Center, University of Alaska Fairbanks, Fairbanks, AK, USA

Part I
Climate Change

Assessment of Climate Change Impact on Watershed Hydrology

Somsubhra Chattopadhyay and Manoj K. Jha

Abstract Evidence of pronounced fluctuation in climate variability has caused numerous impact assessment studies of climate variability and change on watershed hydrology. Several methods of impact assessment have been used over the last decade which basically incorporates atmospheric-ocean circulation-based climate models' projection of changes in meteorological variable into the simulation of land surface hydrological processes. In this study, we have evaluated two methods, frequency perturbation and direct use of data, through forcing of a simulation model with data from a suite of global climate models. Hydrologic response of a typical watershed in Midwest was evaluated for the change in climatic condition. Frequency perturbation method found precipitation decrease by 17 % and reduction in temperature by 0.43 °C on an average annual basis. The changes when applied through the watershed simulation model resulted in 13 % reduction in evapotranspiration (ET) and 25 % reduction in water yield. In contrast, direct method with 1.25 % decrease in precipitation and 0.2 °C decrease in temperature on annual basin found an increase of 1.8 % for ET and 5 % reduction in water yield. Changes in ET and water yield on temporal and spatial scale due to changes in future climate are likely to have severe implications for the water availability. However, more research is needed to evaluate several impact assessment methods for more accurate analysis.

Introduction

Hydrological cycle has been found to be significantly impacted by global warming caused by the climate change in recent times. Intergovernmental Panel for Climate Change (IPCC) reported evidences of strong correlations between the increasing amount and concentration of greenhouse gases and aerosols into the atmosphere and the rising global temperature. The impact of climate change on hydrological processes have been investigated across the globe during the last decades in several

---

S. Chattopadhyay • M.K. Jha (✉)
Computational Science and Engineering and Civil, Architectural and Environmental Engineering Department, North Carolina A&T State University, Greensboro, NC, USA
e-mail: schattop@aggies.ncat.edu; mkjha@ncat.edu

studies which clearly emphasizes the consequence of climatic change and variability on water resources (Jha et al. 2004, 2006, 2010a, b; Jha and Gassman 2013; Jin and Sridhar 2012; Mango et al. 2011). Global climate models (GCMs) are considered as the most important tool to conduct climate change impact assessment studies. Wilby et al. (2004) pointed out that variations in local climate are mainly governed by the regional physiographic conditions which often are not accurately represented by the coarse resolutions of GCM outputs and this further puts into question the process of forcing these information into hydrologic models.

There are several methods available that are used to create future variations in local scale climate from the GCM outputs. Delta change method involves altering the observed temperature and precipitation series according to the "expected" future change signal from the GCMs (Hay et al. 2000; Prudhomme et al. 2002). While temperature has shown good agreement across GCMs according to this method, there has not been the corresponding response for precipitation. In addition, keeping the number of wet days unchanged along with discarding potential changes in correlation among different variables might result into neglecting climate variability. Frequency perturbation change method essentially implies transferring the extracted climate change signals to the observed series which accounts for the changes in extreme rainfall events (Taye et al. 2011; Mora et al. 2013). This approach also provides predictions consistent with the occurrence of wet days and wet day rainfall amounts. Rainfall series is perturbed in relation to their frequency of occurrence. Thus, daily rainfall amounts are perturbed with a unique factor dependent on return period. Direct method (Takle et al. 2005, 2010) uses direct output of GCM runs into the hydrologic simulation and thus takes into account more complex changes in the probability functions of the input weather variables into hydrological models. Bias could be perpetuated in some occasions which accounts as a disadvantage of direct method.

In this chapter, we examined a range of methods and attempted to devise a strategy while clarifying some of the issues with the impact assessment studies. Frequency perturbation method and direct method were applied to obtain watershed-scale future climatic information from a suite of 10 GCMs which were then forced to the hydrologic simulation model Soil and Water Assessment Tool (SWAT). Temporal variations of the major hydrological variables such as evapotranspiration, water yield, surface runoff, and baseflow were evaluated for impact assessment. A modeling framework for a typical Midwestern watershed, developed by Jha et al. (2010a, b), was used to test these two methods.

Materials and Methods

Study area: Raccoon River Watershed covers an area of 9400 km^2 in west-central Iowa before finally draining to larger Des Moines River Watershed. Landuse pattern have been dominated by agricultural crop production mainly corn and soybean (70 %) followed by grassland (16.3 %), woodland (4.4 %), and urban

(4 %) areas. Raccoon River also is of chief importance for the central Iowa region as it serves as a potable water source for nearly 500,000 people.

Hydrological modeling with SWAT: SWAT is a long-term, continuous, watershed-scale simulation model that operates on a daily time step and is designed to assess the impact of different management practices on water, sediment, and agricultural chemical yields. The model is distributed, computationally efficient, and capable of simulating a detailed level of spatial detail (Arnold et al. 1998). It simulates the hydrological cycle based on the water balance equation. Major model components are hydrology, weather, soil temperature, crop growth, nutrient, bacteria, and land management. SWAT divides a watershed into several subwatersheds which then are further delineated according to unique combination of landuse, soil type, and soil class called Hydrologic Response Units (HRUs). HRUs are the smallest possible division of a watershed and the model accounts for the water balance of each subwatershed over the individual HRUs. Each HRU has four distinct components, i.e., snow, soil profile, shallow, and deep aquifer contributing towards total water balance.

Climate change impact assessment methods: The projection simulation was based on A1B Special Report Emission Scenario (SRES) which basically puts a balanced emphasis on all energy sources in future. In this study, we have chosen to compare the frequency perturbation approach with the direct method. Expected changes in rainfall was determined as the ratio of the value in the scenario period to the value of the control period, known as perturbation factor while temperature was changed according to difference between control and scenario period. For rainfall, frequency analysis of quantiles method was applied where perturbation factors were obtained by comparing quantiles for given empirical return periods (or values of the same rank) in both the control and scenario series (Chiew 2006; Harrold et al. 2005; Olsson et al. 2009; Willems 2011; Mora et al. 2013). This perturbation calculation was performed considering only wet days where a wet day was defined as a day receiving a minimum rainfall amount of 0.1 mm. We selected 0.1 mm as a standard wet-day threshold based on previous studies such as in Elshamy et al. (2009). Changes in the wet-day frequencies were also calculated following quantile perturbation calculation for the wet-day rainfall intensities. The day-to-day variability was addressed through the adjustment of the length of wet and dry spells. We have used a random approach that kept altering the wet and dry spells. Wet spell was defined as any span of time longer than two consecutive days receiving more than 0.1 mm rainfall. The change in mean wet spell length was then calculated from the wet spells in the control and scenario GCM runs on a monthly basis and was adjusted in the observed rainfall series through adding or removing wet days to the beginning or end of the wet spells in the series. Thus, we have perturbed observed rainfall series in two steps first by removing or adding wet days in the series using the random approach described earlier and secondly by applying intensity perturbation to each wet day dependent on the empirical return period of the rainfall intensity. Direct method implied running the hydrologic simulation with the BIAS corrected GCM data for both current and future conditions. Contemporary GCM data was used as the baseline scenario while future GCM data was used for the mid-century scenario.

Results and Discussions

Projected climate change in the Raccoon River Watershed in mid-century: Changes in climate occurring over both North and South RRW were analyzed using an ensemble of 10 GCM simulations driven by the A1B of the SRES scenario. Raccoon River Watershed was found to have an average 17 % decrease in monthly precipitation while average temperature is expected to decline by 0.43 °C in the mid-century according to the mean projection of all the GCMs.

Monthly analysis for the watershed (Fig. 1) also suggested that climate could have some interesting variations in future. It was found that for RRW precipitation was mostly decreasing on a monthly basis. Summer months comprising May, June, July, and August showed an average of 16 % decrease while winter months of December, January, and February displayed an 18 % decline. This trend of precipitation over the watershed in mid-century clearly suggests that water scarcity could hamper the agricultural practices during the summer months as crops in the growing season needs more water. This trend in precipitation could also impact the hydrological behavior of the watershed as the water input to the system is substantially reduced in a consistent basis. Projected monthly patterns of temperature for RRW in mid-century were found to have different trend than precipitation. While the climate models suggest mostly an increasing pattern for average daily temperature for winter months of November, December, and January, it is predicted that average daily temperature would be decreasing for the other months of the year (Fig. 2) Winter months evidenced an increase of 0.22 °C in average temperature while summer months showed a decrease of 1.06 °C. On an average, monthly average temperature was found to reduce by 0.42 °C.

Impact on hydrological response: After the future scenarios were developed for rainfall and temperature, original and perturbed series were then used to drive the hydrological model to assess the influence of climate change. Hydrological simulation results were statistically processed to study the impact of climate change.

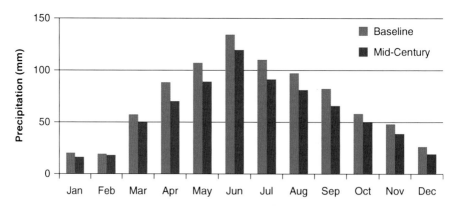

Fig. 1 Comparison of precipitation for baseline (1983–2000) and perturbed baseline according to the frequency perturbation method for the mid-century (2046–2063)

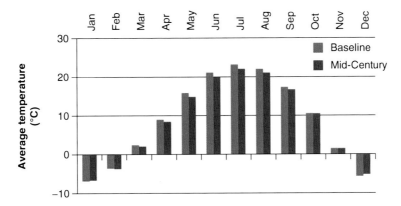

Fig. 2 Comparison of average temperature for baseline (1983–2000) and perturbed baseline according to absolute change method for the mid-century (2046–2063)

Analysis of the changes in terms of both magnitude and percentage suggest that chief hydrological components were significantly affected by the climatic conditions of mid-century. Annual average decrease of 17 % precipitation along with a decrease of 0.43 °C average temperature produced significant changes on projected ET, water yield, and thus overall hydrologic balance. Changes in surface runoff and water yield were found to be 48 % and 25 %, respectively, while baseflow was found to be reduced by 8 %. ET was found to decline by 13 % in mid-century with a magnitude of 79 mm on an annual average basis. A decrease in ET is primarily caused by both reductions in temperature and precipitation. Due to highly nonlinear and complex interactions between the different components of water movement, changes in surface runoff and water yield are not proportional. Water yield which is the total amount of water available at any time was found to decrease by 60 mm while surface runoff and baseflow reduced by 42 and 11 mm, respectively. Monthly variations of water yield (Fig. 3a) showed a wide range of reduction varying from 8 % in August to 62 % in April. April was the highest impact month in terms of reduction in water yield. Winter months are expected to be more affected in terms of water yield reduction with 34 % decline from baseline than summer months when the reduction was found to be 19 %. On an average, 30 % reduction was noticed in water yield on a monthly basis which clearly implies RRW might suffer from water scarcity in the mid-century. ET showed (Fig. 3b) almost a clear decreasing trend for all the months except April when it increased by 34 %. Maximum reduction of water yield was found in the month of April and the increase in ET could be attributed as a reason. A 16 % decrease in precipitation in the summer months almost produced proportional results in ET and water yield as they declined by 19 % and 20 %, respectively. Reason behind decreasing trend of ET is most likely due to the decreasing pattern of the two main variables governing it, i.e., precipitation and temperature. On an average, ET was found to be decreasing from baseline conditions by 15 % on a monthly basis.

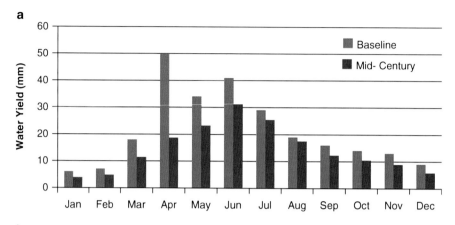

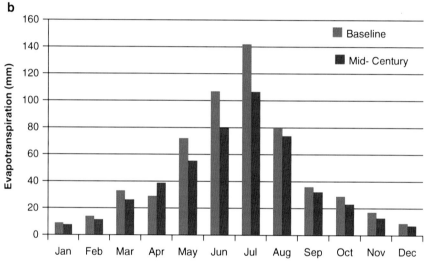

Fig. 3 Comparison of (**a**) water yield and (**b**) evapotranspiration as simulated by SWAT for the baseline and mid-century according to frequency perturbation method

Hydrologic balance of water movement is controlled by precipitation, and evapotranspiration, which finally results in water yield. Directions of water yield changes are following the same trend as precipitation although the changes for water yield are greater in magnitude than the precipitation changes. These results proved to agree with GCM precipitation and temperature predictions used in the hydrologic simulation.

Direct method analysis: Climate change impact assessment was also done using the direct method which assumes that the GCMs are simulating the weather pattern well. It should be mentioned that for direct method "baseline" corresponded to the hydrologic simulation using climate model contemporary data while mid-century corresponded to hydrologic simulation using the future climatic data from the

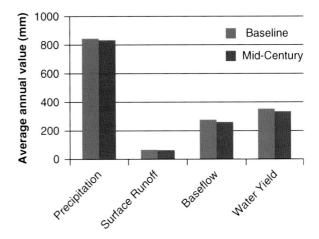

Fig. 4 Comparison of precipitation over the baseline and mid-century scenario for the Raccoon River watershed according to the direct method

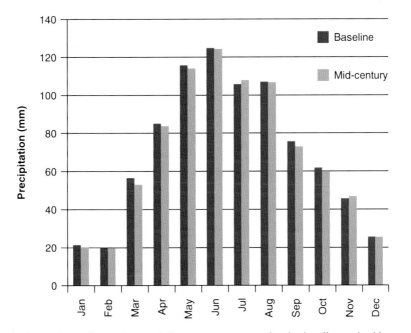

Fig. 5 Comparison of annual water balance components under the baseline and mid-century conditions according to the direct method

climate models. It was found that on an average annual basis, precipitation got reduced by 1.25 % in mid-century as compared to the baseline climatic conditions. Annual average temperature got reduced by 0.17 °C in future. Monthly variations of precipitation revealed mostly decreasing pattern with the exceptions of February, July, and November (Fig. 4). Figure 5 displays the annual water balance components for current and future conditions. ET displayed an increase of 1.8 % on an average annual basis while water yield got reduced by 5 %. Possible interpretation of decreasing water yield could be reducing surface runoff and baseflow by 1.87 % and 6 %, respectively, from the baseline scenario. Comparing the two methods of impact assessment shows an interesting pattern for monthly precipitation. While direct method displayed reductions in small range for winter (December, January, and February) and slight increase for summer months, frequency perturbation method showed steady decrease over all the seasons with the most in winter. However, the general trend appears to be that of a reducing nature.

Conclusions

This study looked upon climate change impact assessment using frequency perturbation approach and direct method. Both methods agreed that watershed is expected to receive less rainfall and the temperature is also expected to reduce in mid-century. For frequency perturbation approach, precipitation declined by 17 % while average temperature reduced by 0.43 °C. Direct method produced 1.25 % decrease in precipitation in mid-century while average annual temperature got reduced by 0.17 °C. According to the frequency perturbation method, ET was found to reduce by 13 %, water yield decreased by 25 % while direct method showed an increase of 1.8 % for ET on an average annual basis while water yield got reduced by 5 %. It was inferred that water scarcity could be an alarming issue for this watershed. By performing the detailed analysis using these two methods, it can be concluded that water resources need to be managed in an efficient way in near future for this region particularly from the agricultural production perspective. Further research is also anticipated considering other downscaling methods of GCM projections on this watershed to have even broader range of climate change impact assessment.

References

Arnold, J. G., Srinivasan, R., Muttiah, R. S., & Williams, J. R. (1998). Large area hydrologic modeling and assessment part I: Model development. *J. Amer. Water Resour. Assoc. 34*(1): 73–89.
Elshamy, M. E., Sayed, M. A.-A. & Badawy, B. (2009). Impacts of climate change on Nile flows at Dongola using statistically downscaled GCM scenarios. Nile Water Science & Engineering Magazine, 2: 1–14. Special issue on Water and Climate.
Chiew, F. H. S. (2006). An overview of methods for estimating climate change impact on runoff. In *30th Hydrology and Water Resources Symposium*, Lauceston, Australia, pp. CDROM (ISBN 0-8582579-0-4).
Harrold, T. I., Chiew, F. H. S., & Siriwardena, L. (2005). A method for estimating climate change impacts on mean and extreme rainfall and runoff. In *16th International Congress on Modelling and Simulation*, Melbourne, Australia, pp. 497–504.

Hay, L. E., Wilby, R. L., & Leavesly, G. H. (2000). A comparison of delta change and downscaling GCM scenarios for three mountainous basins in the United States. *Journal of the American Water Resources Association, 36*, 387–398.

Jha, M. K., Arnold, J. G., Gassman, P. W., Giorgi, F., & Gu, R. (2006). Climate change sensitivity assessment on Upper Mississippi River Basin streamflows using SWAT. *Journal of the American Water Resources Association, 42*(4), 997–1016.

Jha, M. K., & Gassman, P. W. (2013). Changes in hydrology and streamflow as predicted by modeling experiment forced with climate models. *Hydrological Processes, 28*(5), 2772–2781. doi:10.1002/hyp.9836.

Jha, M. K., Pan, Z., Takle, E. S., & Gu, R. (2004). Impact of climate change on stream flow in the Upper Mississippi River Basin: A regional climate model perspective. *Journal of Geophysical Research, 109*, D09102. doi:10.1029/2003JD003686.

Jha, M. K., Schilling, K. E., Gassman, P. W., & Wolter, C. F. (2010a). Targeting landuse change for nitrate-nitrogen load reductions in an agricultural watershed. *Journal of Soil and Water Conservation, 65*(6), 342–352. doi:10.2489/jswc.65.6.342.

Jha, M. K., Wolter, C. F., Schilling, K. E., & Gassman, P. W. (2010b). Assessment of total maximum daily load implementation strategies for nitrate impairment of the Raccoon River, Iowa. *Journal of Environmental Quality, 39*, 1317–1327. doi:10.2134/jeq2009.0392.

Jin, X., & Sridhar, V. (2012). Impacts of climate change on hydrology and water resources in the Boise and Spokane River Basins. *Journal of the American Water Resources Association, 48*(2), 197–220. doi:10.1111/ j.1752-1688.2011.00605.x.

Mango, L. M., Melesse, A. M., McClain, M. E., Gann, D., & Setegn, S. G. (2011). Hydrometeorology and water budget of the Mara River Basin under land use change scenarios. In A. M. Melesse (Ed.), *Nile River Basin: Hydrology, climate and water use* (pp. 39–68). Dordrecht, The Netherlands: Springer.

Mora, D. E., Campozano, L., Cisneros, F., Wyseure, G., & Willems, P. (2013). Climate changes of hydrometeorological and hydrological extremes in the Paute basin. *Ecuadorean Andes Hydrology and Earth System Science Discussions, 10*, 6445–6471.

Olsson, J., Berggren, K., Olofsson, M., & Viklander, M. (2009). Applying climate model precipitation scenarios for urban hydrological assessment: A case study in Kalmar City, Sweden. *Atmospheric Research, 92*, 364–375. doi:10.1016/j.atmosres.2009.01.015.

Prudhomme, C., Reynard, N., & Crooks, S. (2002). Downscaling of global climate models for flood frequency analysis: Where are we now? *Hydrological Processes, 16*, 1137–1150.

Takle, E. S., Jha, M. K., Lu, E., Arritt, R. W., Gutowski, W. J., Jr., & the NARCAP Team. (2010). Streamflow in the Upper Mississippi River Basin as simulated by SWAT driven by 20th century contemporary results of global climate models and NARCCAP regional climate models. *Meteorologische Zeitschrift, 19*(4), 341–346. doi:10.1127/0941-2948/2010/0464.

Takle, E. S., Jha, M. K., & Anderson, C. J. (2005). Hydrological cycle in the Upper Mississippi River Basin: 20th century simulations by multiple GCMs. *Geophysical Research Letters, 32*, L18407. doi:10.1029/2005GL023630.

Taye, M. T., Ntegeka, V., Ogiramoi, N., & Willems, P. (2011). Assessment of climate change impact on hydrological extremes in two source regions of the Nile River Basin. *Hydrology and Earth System Science, 15*, 209–222.

Wilby, R. L., Charles, S. P., Zorita, E., Timbal, B., Whetton, P., & Mearns, L. O. (2004). Guidelines for use of climate scenarios developed from statistical downscaling methods. Supporting material of the Intergovernmental Panel on Climate Change. Available from the DDC of IPCC TGCIA, 27.

Willems, P. (2011). VHM approach: transparent, step-wise and data mining based identification and calibration of parsimonious lumped conceptual rainfall-runoff models. *Journal of Hydrology*, revised.

Trend of Climate Variability in North Carolina During the Past Decades

Mohammad Sayemuzzaman and Manoj K. Jha

Abstract Trend of climate variability in North Carolina for the period of 1950–2009 was investigated in this study with annual scale minimum temperature (T_{min}), maximum temperature (T_{max}), mean temperature (T_{mean}), and precipitation data series from 249 evenly distributed meteorological stations. The trends were tested using Mann–Kendall (MK) test. Theil–Sen approach (TSA) and Sequential Mann–Kendall (SQMK) test were also applied to detect the magnitude and abrupt change of trend, respectively. Lag-1 serial correlation and double mass curve analysis were adopted to check the independency and in homogeneity of the data sets, respectively. For most regions and over the period of past 60 years, trend of T_{min} was found increasing (on 73% of the stations) while for T_{max}, it was found decreasing (on 74 % of the stations). Although the difference between T_{max} and T_{min} trends were decreasing, but increasing trend in T_{mean} represent the overall temperature increasing pattern in North Carolina. Magnitude of T_{max}, T_{min}, and T_{mean} were found to be −0.05 °C/decade, +0.08 °C/decade, and +0.02 °C/decade, respectively, as determined by the TSA method. The SQMK test identified a significant positive shift of T_{mean} during 1990s. For precipitation trends analysis, almost equal nos. of stations was showing statewide positive and negative trends in annual time series. Annually, positive (negative) significant trends, seven (three) nos. of stations were observed at the 95 and 99 % confidence levels. A magnitude of precipitation trend of +3.3 mm/decade was calculated by the TSA method. No abrupt shift was found in precipitation data series over the period by the SQMK test.

M. Sayemuzzaman
Energy and Environmental System Department, North Carolina A&T State University, Greensboro, NC, USA

M.K. Jha (✉)
Department of Civil, Architectural and Environmental Engineering,
North Carolina A&T State University, Greensboro, NC, USA
e-mail: mkjha@ncat.edu

Introduction

Historical trends in surface climate components (such as temperature and precipitation) have received considerable attention in recent years. To predict climate shift, floods, droughts, loss of biodiversity and agricultural productivity, changes in temperature and precipitation pattern needs to be analyzed.

In temperature trend analysis, Degaetano and Allen (2002) found a significant increase in high temperature (both maximum and minimum) across the United States from 1910 to 1996, particularly at urban sites. Trenberth et al. (2007) concluded that the south-eastern United States is one of the few regions on this planet showing cooling trend during the twentieth century. In precipitation analysis, Karl and Knight (1998) reported a 10 % increase in annual precipitation across the United States between 1910 and 1996. Total precipitation has increased across the United States over the last several decades as also found by recent research (Kunkel et al. 2002; Small et al. 2006).

Local climate variability is always more accurate/preferable than global or continental scales because of the finer resolution data (Trajkovic and Kolakovic 2009; Martinez et al. 2012). North Carolina has diverse topographic zone from west mountainous region to east coastal region. The nature of the topography makes complex weather pattern in North Carolina (Robinson 2005). There has been very few published works so far found on climatic variables pattern analysis in North Carolina. Boyles and Raman (2003) predicted precipitation and temperature trend in North Carolina on seasonal and annual time scales during the period of 1949–1998 utilizing 75 precipitation measuring stations. Linear time series slopes were analyzed to investigate the spatial and temporal trends of precipitation. They found that temperatures in North Carolina during the 1950s are the warmest in 50 years, but the last 10 years are warmer than average. They predicted that the precipitation of last 10 years in the study period was the wettest. They also found that precipitation has increased over the past 50 years during the fall and winter seasons, but decreased during the summer.

For the time series trend analysis and the shift of trend detection in hydro-meteorological variables, various statistical methods have been developed over the years (Modarres and Sarhadi 2009; Tabari et al. 2011; Martinez et al. 2012; Sonali and Nagesh 2013). Nonparametric method has been favored over parametric methods (Sonali and Nagesh 2013). The chapter presents the trend analysis results for T_{max}, T_{min}, T_{mean}, and average precipitation on an annual scale utilizing the time series data from 249 weather stations across the state of North Carolina.

Materials and Methods

Study area and data. Total area of North Carolina is 52,664 mi^2 which is situated in the southeastern United States (34°–36° 21′ N and 75° 30′–84° 15′ W). Daily T_{max}, T_{min}, and precipitation data series of 249 stations were collected from the United

States Department of Agriculture-Agriculture Research Service (USDA-ARS 2012). The data sets facilitated and quality controlled by National Oceanic and Atmospheric Administration (NOAA) which includes the meteorological stations of both Cooperative Observer network (COOP) and Weather-Bureau-Army-Navy (WBAN). We consider the data sets of the stations based on record length, record completeness, spatial coverage, and historical stability over the period 1950–2009. The data sets are 99.99 % complete (USDA-ARS 2012). Though these data sets have been quality controlled, the double mass curve employed to detect the non-homogeneity/inconsistencies of the data sets, if any. Instrument changes, station shifts, changes of land cover/surrounding conditions may create the non-homogeneity/inconsistencies in hydro-meteorological data recording (Tabari et al. 2011). Tabari and Hosseinzadeh Talaee (2011) applied the double mass curve on their climate variables data sets to check the inconsistency. In our double mass curve analysis, we found almost a straight line with no obvious break points at all the stations of T_{max}, T_{min}, and precipitation data series within the study period. In this study, dense (1 per 548 sq. km.) observation station data indicate an important component of the analyses.

Seasonal and annual time series were obtained from the averaging of daily data for each of the 249 stations. Seasons as adopted from Boyles and Raman (2003) can be defined as follows: Winter (January, February, March); spring (April, May, June); summer (July, August, September); and fall (October, November, December).

Trend Analysis

Mann–Kendall test. The Mann–Kendall test is one of the widely used nonparametric tests to detect significant trends in hydro-meteorological time series. This test makes the comparison of the relative magnitudes of the sample data rather than the data values itself. The most salient features of the MK test are: (a) possess low sensitivity for the nonhomogeneous/inconsistent data sets (Modarres and Sarhadi 2009) and (b) doesn't require the data sets to follow any particular distribution (Gocic and Trajkovic 2013).

Theil–Sen approach (TSA). The MK test does not provide an estimate of the magnitude of the trend. For this purpose in this study, a nonparametric method referred to as the Theil–Sen approach (TSA) is used. TSA approach is originally described by Theil (1950) and Sen (1968). This approach provides a more robust slope estimate than the least-squares method because it is insensitive to outliers or extreme values and competes well against simple least squares even for normally distributed data in the time series (Jianqing and Qiwei 2003). TSA approach is also known as Sen's slope estimator. Sen's slope estimator has been widely used by researchers for the trend magnitude prediction in hydro-meteorological time series (Tabari et al. 2011; Martinez et al. 2012).

Sequential Mann–Kendall (SQMK) test. SQMK test is an extension of the MK method which is widely used to detect the time when trend has a shift (Modarres and Sarhadi 2009; Sonali and Nagesh 2013). SQMK is a sequential forward ($u(t)$) and backward ($u'(t)$) analyses of the MK test. If the two series are crossing each other, the year of crossing exhibit the year of trend change (Modarres and Sarhadi 2009). If the two series crosses and diverge to each other for a longer period of time, beginning diverge year exhibit the abrupt trend change (Tabari et al. 2011).

Pre-whitening. The presence of serial correlation in the time series makes the frequent rejection of the null hypothesis in the MK test, especially if there is any positive serial correlation (Kulkarni and von Storch 1995; von Storch 1995). Thus, both MK test and TSA approach require time series to be serially independent which can be accomplished by using the pre-whitening technique suggested by von Storch and Navarra (1995).

Results and Discussion

In Fig. 1, T_{max} (T_{min}) shows the higher negative (positive) trends in 74 % (73 %) of stations. From these percentages of trends, 13 % (19 %) of stations were showing significant (confidence level ≥95 %) negative (positive) trends in T_{max} (T_{min}) data series. Precipitation data series were showing 52 % stations of positive and 48 % stations of negative trends with significant (confidence level ≥95 %) negative (positive) trends of 2 % (3 %) stations.

In Fig. 2, the time series, along with the 5-year moving average and the trend line (from the TSA) average of 249 stations, were graphically represented for (a) T_{max}, (b) T_{min}, (c) T_{mean}, and (d) precipitation data series over the period 1950–2009. Overall, trend of T_{max} is decreasing and T_{min} is increasing. These results are consistent with the earlier studies of Boyles and Raman (2003). Statewide mild precipitation increasing trend was noticed. Magnitude of T_{max}, T_{min}, T_{mean}, and precipitation trends were found −0.05 °C/decade, 0.08 °C/decade, +0.02 °C/decade, and +3.3 mm/decade, respectively.

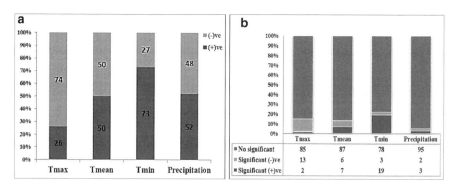

Fig. 1 Percentage of stations (in 249 stations) with (**a**) positive–negative trends, (**b**) significant (confidence level ≥95 %) positive–negative trends determined by Mann–Kendall test across North Carolina over the period 1950–2009

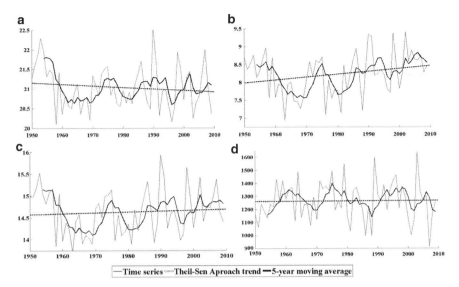

Fig. 2 Observed, 5-year moving average and trend line of annual (**a**) T_{max}, (**b**) T_{min}, (**c**) T_{mean}, and (**d**) precipitation data series average of all the stations (249 nos.) across North Carolina over the period 1950–2009

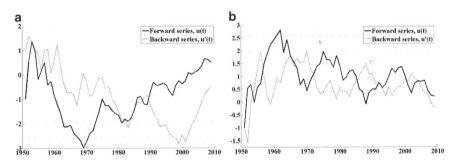

Fig. 3 Graphical representation of the forward-backward series of Sequential Mann–Kendall (SQMK) test of (**a**) T_{mean} and (**b**) precipitation data series average of all the stations (249 nos.) across North Carolina over the period 1950–2009

SQMK test was applied to the mean temperature and precipitation data series of average of all the stations (249 nos.) across North Carolina over the period 1950–2009 to detect the beginning year of trend or change (Fig. 3). In Fig. 3a, after 1990, the mean temperature increasing trend shift was noticeable. In Fig. 3b, precipitation data series was not showing any abrupt change.

Conclusion

In this study, annual time series of long-term historic (1950–2009) climatic variables (temperature and precipitation) trends were analyzed. This chapter develops for the first time a full picture of temperature and precipitation trends with the denser observation stations (1 per 548 sq. km) totaling 249 across the state of North Carolina. Further research is needed to identify the driving forces of these trends.

References

Boyles, R. P., & Raman, S. (2003). Analysis of climate trends in North Carolina (1949–1998). *Environment International, 29*, 263–275.

Degaetano, T. A., & Allen, J. R. (2002). Trends in twentieth-century temperature extremes across the United States. *Journal of Climate, 15*, 3188–3205.

Gocic, M., & Trajkovic, S. (2013). Analysis of precipitation and drought data in Serbia over the period 1980–2010. *Journal of Hydrology, 494*, 32–42. doi:10.1016/j.jhydrol.2013.04.044.

Jianqing, F., & Qiwei, Y. (2003). Nonlinear time series: Nonparametric and parametric methods. Springer series in statistics. New York: Springer. ISBN 0387224327.

Karl, T. R., & Knight, R. W. (1998). Secular trend of precipitation amount, frequency, and intensity in the United States. *Bulletin of the American Meteorological Society, 79*, 231–242.

Kulkarni, A., & von Storch, H. (1995). Monte-Carlo experiments on the effect of serial correlation on the Mann-Kendall test of trend. *Meteorologische Zeitschrift, 4*(2), 82–85.

Kunkel, K., Andsager, K., Liang, X., Arritt, R., Takle, E., Gutowski, W., et al. (2002). Observations and regional climate model simulations of heavy precipitation events and seasonal anomalies: A comparison. *Journal of Hydrometeorology, 3*, 322–334.

Martinez, J. C., Maleski, J. J., & Miller, F. M. (2012). Trends in precipitation and temperature in Florida, USA. *Journal of Hydrology, 452–453*, 259–281.

Modarres, R., & Sarhadi, A. (2009). Rainfall trends analysis of Iran in the last half of the twentieth century. *Journal of Geophysical Research, 114*, D03101. doi:10.1029/2008JD010707.

Robinson, P. (2005). North Carolina weather and climate. University of North Carolina press in association with the State Climate Office of North Carolina (Ryan Boyles, graphics).

Sen, P. K. (1968). Estimates of the regression coefficient based on Kendall's tau. *Journal of American Statistics Association, 63*(324), 1379–1389.

Small, D., Islam, S., & Vogel, R. M. (2006). Trends in precipitation and streamflow in the eastern US: Paradox or perception? *Geophysical Research Letter, 33*, L03403. doi:10.1029/2005GL024995.

Sonali, P., & Nagesh, K. D. (2013). Review of trend detection methods and their application to detect temperature changes in India. *Journal of Hydrology, 476*, 212–227.

Tabari, H., & Hosseinzadeh Talaee, P. (2011). Analysis of trends in temperature data in arid and semi-arid regions of Iran. *Global and Planetary Change, 79*(1–2), 1–10.

Tabari, H., Shifteh, S. B., & Rezaeian, Z. M. (2011). Testing for long-term trends in climatic variables in Iran. *Atmospheric Research, 100*(1), 132–140.

Theil, H. (1950). A rank-invariant method of linear and polynomial regression analysis. *Nederlandse Akademie Wetenschappen Proceedings, A53*, 386–392.

Trajkovic, S., & Kolakovic, S. (2009). Wind-adjusted Turc equation for estimating reference evapotranspiration at humid European locations. *Hydrology Research, 40*(1), 45–52.

Trenberth, K. E., Jones, P. D., Ambenje, P., Bojariu, R., Easterling, D., Klein Tank, A., et al. (2007). Observations: Surface and atmospheric climate change. In S. Solomon, D. Qin, M. Manning, Z. Chen, M. Marquis, K. B. Averyt, et al. (Eds.), *Climate change 2007: The physical science basis. Contribution of working group I to the fourth assessment report of the intergovernmental panel on climate change*. Cambridge, England: Cambridge University Press.

USDA-ARS. (2012). Agricultural Research Service, United States Department of Agriculture. Retrieved November, 2012, from http://www.ars.usda.gov/Research/docs.htm?docid=19440.

von Storch, H. (1995). Misuses of statistical analysis in climate research. In H. V. Storch & A. Navarra (Eds.), *Analysis of climate variability: Applications of statistical techniques* (pp. 11–26). Berlin, Germany: Springer.

von Storch, H., & Navarra, A. (1995). *Analysis of climate variability—Applications of statistical techniques*. New York: Springer.

Arctic Storm and Its Impact on the Surface Winds over the Chukchi-Beaufort Seas

Wei Tao, Jing Zhang, Xiangdong Zhang, and Junming Chen

Abstract Storms generating in the Chukchi-Beaufort Seas or traveling to the area play an important role impacting the regional climate and environment. Surface wind associated with storm exerts dynamic forcing, modifying sea ice movements and ocean currents, opening sea ice leads, and redistributing sea ice coverage. In this study, Arctic storms and their impacts on the surface wind field in the Chukchi-Beaufort Seas are studied with the newly developed Chukchi-Beaufort High-resolution Atmospheric Reanalysis (CBHAR) for the period of 1979–2009. Storm activities over the Chukchi-Beaufort Seas region show a seasonal variability. Storms are less but stronger in cold season, and more in number but weak in warm season. The storms impact the area's surface wind field by enhancing the frequency of strong winds. When the storms interact with the semi-permanent weather system Beaufort High, extremely strong winds often occur.

Introduction

The Chukchi-Beaufort Seas region of the Arctic is currently undergoing significant environmental changes, including the fastest rate of decline and maximum observed interannual variance of sea ice anywhere in the Arctic (Comiso 2012), along with increased surface wind speeds over recent decades as the sea ice retreats (Stegall and Zhang 2012). In addition to the natural changes, the potential for further offshore energy development also exists in the region. With oil extraction comes the threat of oil spills, which can have serious environmental consequences. Oil spills may impact not only the immediate area but also remote regions due to transport via ocean currents and drifting sea ice. Given that the Chukchi–Beaufort Seas comprise a particularly vulnerable and fragile region, with an ecosystem and

W. Tao (✉) • J. Zhang
Department of Physics and EES, North Carolina A&T State University, Greensboro, NC 27411, USA
e-mail: taoweijanet@gmail.com

X. Zhang • J. Chen
International Arctic Research Center, University of Alaska Fairbanks, Fairbanks, AK 99775, USA

© Springer International Publishing Switzerland 2016
G.A. Uzochukwu et al. (eds.), *Proceedings of the 2013 National Conference on Advances in Environmental Science and Technology*,
DOI 10.1007/978-3-319-19923-8_3

environment that are especially sensitive to human impacts, it's critical to have a good understanding of the area's surface wind field, a crucial parameter for assessing and predicting spill transport (Reed et al. 1999).

In order to better understand how the prevailing weather system interacts with local finer-scale processes and how these in turn impact the surface wind field over the Chukchi–Beaufort Seas region, the Weather Research and Forecasting (WRF) model (Skamarock et al. 2008) and its data assimilation system WRFDA (Huang et al. 2009; Barker et al. 2012) were utilized to generate a long-term, high-resolution regional reanalysis over the study area. The outcome of this effort is the 31-year Chukchi–Beaufort High-Resolution Atmospheric Reanalysis (CBHAR) (Zhang et al. 2013).

Storms generating in the Chukchi-Beaufort Seas or traveling to the area impact the regional climate and environment and play a fundamental contributing role in transporting atmospheric heat energy and moisture into the Arctic (Simmonds et al. 2008; Zhang et al. 2004). Surface winds associated with storms exerts dynamic forcing, modifying sea ice movements and ocean currents, opening sea ice leads, and redistributing sea ice cover (Long and Perrie 2012; Simmonds and Keay 2009). Furthermore, the extreme winds induced by the intense storms can give rise to high wave surges, causing flooding, coastal erosion, and property damage, particularly after the retreat of sea ice (Lynch et al. 2004; Small et al. 2011).

In this study, based on the newly developed CBHAR, the climatology of storm activity in the Chukchi-Beaufort Seas region was studied, and the storm impacts on the area's surface wind field will be investigated. The CBHAR data and storm identification algorithm used in this study are described in section "CBHAR Reanalysis and Storm Identification Algorithm". The climatological circulation in the Chukchi-Beaufort Seas region is presented in section "Climatological Circulation in the Chukchi-Beaufort Seas Region". Storm climatology and variability and the impacts on the area's surface wind field are discussed in sections "Storm Climatology and Variability in CBHAR" and "Storm Impacts on the Surface Winds in the Chukchi-Beaufort Seas". Conclusions of this study are given in section "Concluding Remarks".

CBHAR Reanalysis and Storm Identification Algorithm

The data used in this study is CBHAR, which is a three-dimensional, thermodynamically and dynamically constrained gridded dataset at a high resolution of 10 km in horizontal spacing and 1 h time interval over the period 1979–2009 in the Chukchi-Beaufort Seas region (Fig. 1). CBHAR is generated by the WRF model and its data assimilation system WRFDA forced by the European Centre for Medium-Range Weather Forecasts Interim Reanalysis (ERA-I) (Dee et al. 2011). The WRF model physics used in generating CBHAR include the RRTMG (Rapid Radiative Transfer Model for GCMs) radiation (Iacono et al. 2008), the Grell-3D

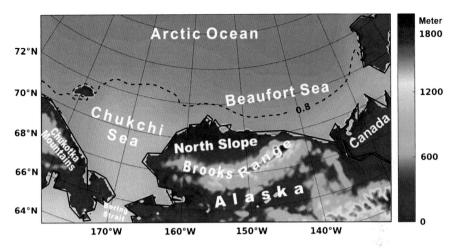

Fig. 1 Study area, encompassing the Chukchi-Beaufort Seas, Chukotka Mountains, Arctic (North) Slope, Brooks Range, and parts of Interior Alaska and the Canadian Yukon. The mean climatological extent of 80 % sea ice concentration is highlighted by the dashed line; topographic height (m) over land is *shaded*

ensemble cumulus (Grell and Devenyi 2002), the Morrison 2-moment microphysics (Morrison et al. 2009), and the MYJ (Mellor–Yamada–Janjic, Mellor and Yamada 1982; Janjic 2002) boundary schemes. In addition, a thermodynamic sea ice model (Zhang and Zhang 2001) is coupled with the NOAH land surface model (Chen and Dudhia 2001) within WRF in order to accurately model the thermal conditions over sea ice. Data assimilated by WRFDA include the in situ surface and radiosonde measurements, the QuikSCAT Sea winds, and the MODIS-retrieved temperature and humidity profiles under clear-sky and snow-free conditions (Liu et al. 2013).

The storm identification and tracking algorithm developed by Zhang et al. (2004) was employed to investigate the storm activity in CBHAR. In order to accurately identify synoptic-scale storms with the parameter of sea level pressure (SLP) in the high spatial and temporal resolution dataset, the algorithm, originally developed for use with the coarse resolution global reanalysis, is modified. The modified criteria used in detecting storm center and track include:

1. If the SLP at one grid point is lower than at the surrounding grid points within 50 km, a cyclone candidate is identified with its center at this grid point. To minimize the impacts of thermodynamically and dynamically induced high-resolution features on the identification of synoptic-scale storms, the SLP field was initially smoothed at each grid point using the 25 surrounding grid points.
2. The minimum pressure gradient calculated between the candidate storm center and grid points 50 km away must be larger than 0.05 hPa $(100$ km$)^{-1}$.
3. The SLP gradient between the central point and each of the four surrounding points both at a radius of 50 and 100 km are compared, and at least three of them must be negative inward (i.e., the central SLP is lower than the surrounding points).

The criterion requiring three of the four to be negative, not all, is created primarily to allow the inclusion of some storms that have open SLP contours directed away from the candidate center.

4. If two candidate storms simultaneously appear within 600 km of one another, they are considered to be part of the same storm system. The original algorithm sets this limit to be 1200 km, but due to the region's high latitude and complex surface forcing, CBHAR has shown that two distinct storms can indeed exist within this distance. The limit was therefore reduced in order to more accurately identify dynamically separate, yet closely positioned storm centers.
5. The storm has a minimum lifetime of 12 h. In addition, if the location of a storm is within 200 km of a storm center identified 1 h earlier, the center is considered to have either moved or reformed from its previous location. Otherwise, a new storm is considered to generate.

In addition to the criteria defined by Zhang et al. (2004), but modified for this study, the following two additional criteria are newly added specifically for the high-resolution data over complex terrain:

1. The candidate storm's central SLP must be less than 1005 hPa.
2. The distance traversed during the storm's entire lifetime must be larger than 250 km.

These two criteria are designed to remove any dynamic and thermodynamic lows forced by the complex terrain.

Climatological Circulation in the Chukchi-Beaufort Seas Region

Synoptic-scale storms are interactively steered by the atmospheric general circulation pattern. In order to better understand the storm climatology, the circulation patterns covering the Beaufort-Chukchi Seas region are presented here. Regional climatological circulation patterns are generally shaped by the influence of the two semi-permanent systems in the area: Beaufort High in the north and Aleutian Low in the south. Therefore, climatologically, higher pressure appears in the north of the study domain and lower pressure in the south throughout the course of the year (Fig. 2). Due to the presence of Beaufort High, surface pressure increases over the Chukchi-Beaufort Seas from January to May. Lower pressure becomes more dominant since from September as the Aleutian Low gradually strengthens.

To further examine the seasonal cycle of the surface pressure as governed by Beaufort High and Aleutian Low, the study domain is divided into northern and southern halves by the blue line shown in Fig. 2. The SLP exhibits a similar increase from January to March over both the northern and southern sub-domains (Fig. 3). This may suggest an increased influence of the intensifying Beaufort High and a decreased influence of the weakening Aleutian Low during this time period.

Arctic Storm and Its Impact on the Surface Winds over the Chukchi-Beaufort Seas 25

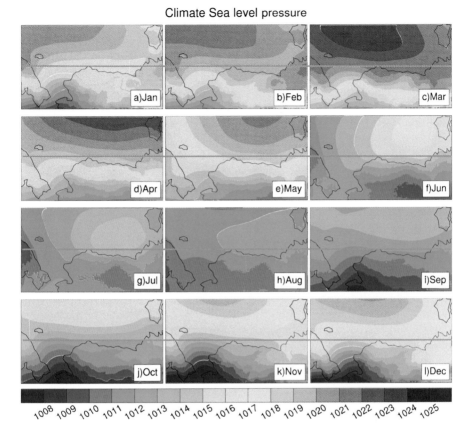

Fig. 2 Climatological monthly mean sea level pressure (SLP, hPa) in CBHAR over 1979–2009. The *blue line* indicates the boundary between the northern and southern sub-domains used for the seasonality analysis

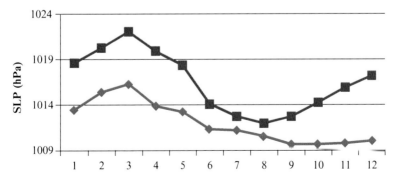

Fig. 3 Climatological monthly mean SLP (hPa) over the northern (*red*) and southern (*blue*) sub-domains in CBHAR during 1979–2009

The intensity of Beaufort High reaches its maximum in March and decreases afterward. In conjunction with this, SLP decreases over the study domain from March to August. In particular, the rate of SLP decrease accelerates in the northern part of the domain from May to August, ending up about 4 hPa lower in August than in May. Note that, during June–August, the difference in the areal mean of SLP between the two subregions declines. During this period, both Beaufort High and Aleutian Low are in their weaker phase. Beginning in September, both Beaufort High and Aleutian Low start to intensify. The SLP correspondingly increases in the north, but, by contrast, remains relatively constant in the south until December. The slight southward shift of the Aleutian Low during December to January favors the increase in SLP over the study domain at this time (Fig. 2). Such seasonal cycle is consistent with the climatology of Beaufort Sea identified by Overland (2009).

Storm Climatology and Variability in CBHAR

Following Zhang et al. (2004), four parameters are employed to describe the storm climatology, including the number, duration, and intensity of storms, as well as an integrative index, called the cyclone activity index (CAI), which measures the overall storm activity in the study region. The analysis is conducted on a monthly basis. The number of storms is defined as the number of storm trajectories located within the study domain for each month; the duration is the mean time period that all counted storms spend within the study domain for each month; and the intensity is the mean difference between the central SLP of all counted storms and the climatological monthly mean SLP at the corresponding grid points throughout the duration of each storm for each month. The CAI is the sum of the differences between the storm's central SLP and the climatological monthly mean SLP at the corresponding grid points for all time steps within the study domain for each month. This index thus integrates all the calculated information about the number, duration, and intensity of the storms.

Applying the modified storm identification and tracking algorithm (as described in section "CBHAR Reanalysis and Storm Identification Algorithm") to the CBHAR data from 1979 to 2009, synoptic storms entering or generated in the Chukchi–Beaufort Seas region are identified. Based on this storm data, the climatology of storm activities in the Chukchi–Beaufort Seas region is presented in Fig. 4. The storm activity demonstrates an obvious seasonal cycle. More numerous, but weaker storms occur in summer, and fewer, but stronger storms appear in the winter season over the study area. This climatological seasonality is precisely consistent with the previous finding for the Arctic Ocean as a whole by Zhang et al. (2004). Considering this contrast between the Arctic seasonality and that of the mid-latitudes, Serreze and Barrett (2008) conducted a follow-up and additionally found a minimum in the number of storms in March. The storm duration displays a much smaller seasonal fluctuation, characterized by a mean duration of about 30 h throughout the year. The seasonal cycle of the CAI indicates an overall

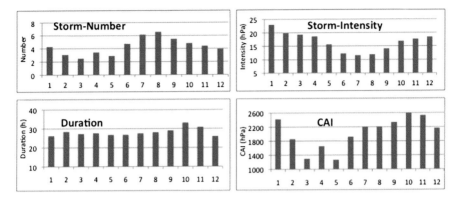

Fig. 4 Climatological monthly mean storm number, intensity (hPa), duration (hours), and CAI (hPa) in CBHAR during 1979–2009

intensification in storm activity from summer to the late winter over the study domain, followed by a weakening in spring. Comparison of the four panels in Fig. 4 suggests that interplay between the number and intensity of storms makes a major contribution to the seasonal variation in the CAI.

The seasonal cycle of storm activity also corresponds well to that of the regional atmospheric circulation patterns. The weakening storm activity during spring is in good agreement with the intensity variations of Beaufort High and Aleutian Low. As discussed above, Beaufort High is strongest from March through May, with its SLP reaching a maximum of 1023 hPa. At the same time, the Aleutian Low weakens noticeably from March onwards (Fig. 2).

Based on the climatological seasonal cycle of storm activity, and considering the seasonal evolution of the regional atmospheric circulation pattern as shown in Fig. 4, the year is divided into four seasons to investigate the year-by-year variability and long-term changes of storm activity. The four seasons are defined as early winter (September–December, when the Aleutian Low grows to its strongest); late winter (January–February, when Beaufort High intensifies and Aleutian Low weakens); spring (March–May, when storm activity is at its weakest); and summer (June–August, when Beaufort High is in its weakest state).

Since CAI is an integrated parameter representing storm activities including number, duration, and intensity, we decide to use CAI investigating the variation of storm activity in the study domain. 31-year seasonal CAI variations demonstrate an obvious interannual variability in all seasons from 1979 to 2009 (Fig. 5). A slight intensification after 1994 appears in the late winter, but there is no significant long-term trend over the entire 31 years. Consistent with the climatological analysis, the interannual variability of storm activity shows seasonal differences, with more storm activities (larger CAI) in both early and late winters, and less in spring. The largest variability occurs in the late winter and the smallest variance in spring.

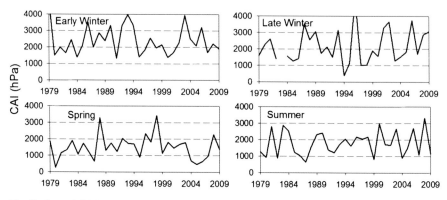

Fig. 5 Seasnal CAI (hPa) variations (Early Winter: September–December; Late Winter: January–February; Spring: March–May; Summer: June–August) in CBHAR during 1979–2009

Storm Impacts on the Surface Winds in the Chukchi-Beaufort Seas

Climatology Aspects

Strong surface winds are always associated with intense storms. To examine how the fluctuation in storm intensity impacts the surface wind field, strong winds, defined as those having speeds exceeding the ninety-fifth percentile, are identified. The ninety-fifth-percentile wind speed is calculated as follows:

1. The speeds are first sorted from smallest to largest at each grid point.
2. The rank R_k is then calculated:

$$R_k = N\frac{P}{100} + 0.5 \qquad (1)$$

where N is the total number of values and P is the percentile (95 here).
3. P_{95} represents the ninety-fifth percentile value calculated by:

$$P_{95} = R_{\text{dec}}(A(R_{\text{whole}} + 1) - A(R_{\text{whole}})) + A(R_{\text{whole}}) \qquad (2)$$

where R_{whole} and R_{dec} are the whole and decimal parts of the rank, respectively as calculated from (2), and A is the sorted array at each grid point.

Among four parameters discussed in section "Storm Climatology and Variability in CBHAR", storm intensity show better correlation with the strong wind. Both the intensity of storms and frequency of strong winds demonstrate large interannual variability across the four seasons (Fig. 6). The two parameters are well correlated in the late winter and summer seasons, with correlation coefficients of 0.46 and 0.40, respectively, which are significant at the 95 % confidence level. This suggests

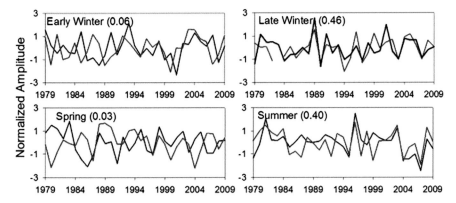

Fig. 6 Standardized time series of storm intensity (*red*) and strong wind (95th-percentile wind speed) frequency (*black*) for each season (Early Winter: September–December; Late Winter: January–February; Spring: March–May; Summer: June–August) in CBHAR. Correlations are given in *parentheses*

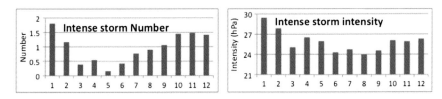

Fig. 7 Climatological number and intensity (hPa) of intense storms for each month from 1979 to 2009 in CBHAR

that the invasion or intensification of storms over the study area may contribute to an increase in surface pressure gradients, in particular when Beaufort High is also present within the domain during late winter or background pressure gradient is calm during summer, and, in turn, increases the frequency of strong winds during these seasons. In contrast, insignificant, small correlations are present for the early winter and spring seasons, in particular before the mid-1990s. This could be attributed to the background cyclone activity associated with Aluetian low in early winter (Fig. 2i–l) dominance of stable Beaufort High in spring (Fig. 2c–e). In general, storms may not be able to play a dominant role in the steering of surface winds.

To further examine the impacts of storms, the analysis will only focus on the intense storms. In particular, a previous study has indicated that a large proportion of strong wind events are related to strong storms (Small et al. 2011). Intense storms are defined here as those with a central SLP lower than 990 hPa. The climatological number and intensity of intense storms in each month are shown in Fig. 7. The seasonal cycle of the number of intense storms exhibits differences from that of all storms. The smallest number of intense storms generally occurs from March to June, with the number increasing after June and reaching its maximum in January.

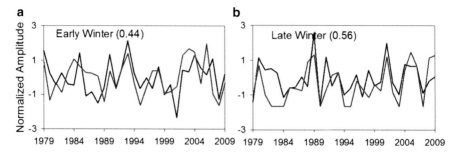

Fig. 8 Standardized intensity of intense storms (*red*) and standardized frequency of strong wind (95th-percentile wind speed) occurrence (*black*) in (**a**) early winter (September–December) and (**b**) late winter (January–February) in CBHAR. Correlations are given in parentheses

The seasonal cycle of intensity has relatively small amplitude, strongest in January and weakest in August. Based on the nature of seasonal cycle, we focus the impact study on surface winds for the early and late winter seasons.

The correlation between the intensity of intense storms and the frequency of strong winds is noticeably larger than the correlation calculated using all storms, in particular for the early winter, which has an increased correlation of 0.44, up from 0.06 (Fig. 8a). These changes, as compared with Fig. 6, suggest that storm in this season should often be intense to generate large pressure gradient in Aluetian low, and become a predominant contributing factor driving the occurrence of strong surface winds. This finding has important implications for assessing the impacts of climate change. Zhang et al. (2004) found that there has been an intensifying trend of storm activity over the pan-Arctic region. If this has and will continue to occur for the current study area, the recently detected and future-projected upward trend in strong surface winds may be attributed to the changes in storm frequency that are the result of a warming climate. To examine this, the yearly frequency of intense storms over the last 31 years is shown in Fig. 8. The results indeed show an increased number of intense storms in the early winter.

Case Studies

From the climatological analysis presented in section "Climatology Aspects", it is found that the variability of intense storms explains about 20–30 % of the total variance of the variability in strong surface winds. Although this value suggests that intense storms have a predominant role in the long-term climate variability of surface winds, other factors can also interact with storms to impact the winds. Considering the regional atmospheric circulation patterns, the interplay between storm and Beaufort High would be the key in shaping strong wind field in the study area, including three possible scenarios:

1. Storms themselves bring strong winds when moving into or being generated within the study domain
2. Intensification of Beaufort High causes strong winds
3. Interaction between storm and Beaufort High generates strong winds

To better understand how the above three scenarios shape the surface wind field, a case study has been conducted for 6–7 October 1992. During this period, an intense storm moved southeastward from Russia over the Bering Strait, reaching its minimum SLP on 6 October 1992 (Fig. 9a). Strong winds occurred over northern Chukchi-Beaufort Seas, with the maximum wind speed exceeding 18 m s^{-1} along the Alaskan coast of the southern Chukchi. Interestingly, a narrow band of strong wind extends from the west coast of Alaska eastward along the Brooks Range, demonstrating topographic effects. The storm continued to maintain its intensity on 7 October 1992 (Fig. 9b), during which time the storm center moved northward

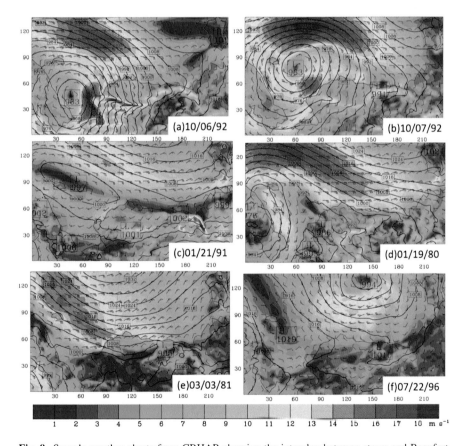

Fig. 9 Sample weather charts from CBHAR showing the interplay between storm and Beaufort High in generating strong surface winds. *Black contours*: SLP (hPa); *shaded colors*: 10-m wind speed (m s^{-1}); *wind barbs*: 10-m winds

over the Chukchi Sea, bringing the associated area of strong winds along with it. At the same time, the strong winds in the extreme southern Chukchi Sea moved northward along the coast. In this case, the intense storm itself played the predominant role in causing the strong surface winds, as a well-shaped Beaufort High had not been formed.

During the late winter, it is very common that storms vigorously interact with a well-established Beaufort High. This can be exemplified by a case occurring on 21 January 1991 (Fig. 9c). The storm was initially situated over the Chukchi Sea with a central SLP of 997 hPa. At the same time, Beaufort High was located to the north with a central SLP higher than 1020 hPa. The placement of storm and Beaufort High resulted in large SLP gradients that generated 12–16 m s^{-1} winds. In this case, the intensity of storm is moderate. In instances when storms are stronger, more extreme winds can be expected. For example, the case occurring on 19 January 1981 shows a storm over the eastern Siberia coastal region with a central SLP of 978 hPa (Fig. 9d). The obviously larger pressure gradients between storm and Beaufort High created strong winds with a maximum wind speed exceeding 18 m s^{-1}.

According to the regional circulation analysis in section "Climatological Circulation in the Chukchi-Beaufort Seas Region", Beaufort High reaches its peak strength in spring, along with Aleutian Low, which is not a fixed feature but rather represents the time-averaged effect of storms traversing the region. Under this circumstance, Beaufort High predominantly governs the formation of surface winds in the area. A case involving strong winds on 3 March 1981 illustrates this effect (Fig. 9e). Here, the central SLP of the Beaufort High was greater than 1032 hPa, and the associated strong winds were around 16 m s^{-1} over the Chukchi Sea.

During summer, Beaufort High is climatologically weak, and as such this season generally has the weakest winds, with a ninety-fifth-percentile wind speed of less than 9 m s^{-1}. However, storms that either move over or are generated over the Arctic Ocean can produce relatively large winds. For example, a storm occurred over the northern Beaufort Sea on 22 July 1996 (Fig. 9f) had a central SLP of 994 hPa, and the associated strong wind speeds were around 9–14 m s^{-1}.

Concluding Remarks

With the newly developed high-resolution regional reanalysis CBHAR, storm activity and its impacts on the surface winds over the Chukchi-Beaufort and Chukchi Seas region are investigated. The storm identification and tracking algorithm originally developed for global reanalysis is modified for detecting storm centers and tracks in the high-resolution regional reanalysis CBHAR. The modified storm identification algorithm successfully captures synoptic storms entering or generated in the Chukchi–Beaufort Seas region over the period 1979–2009. Based on the identified storms, the seasonal cycle and interannual variability of storm numbers and intensity are first analyzed. Then the relationships between storm intensity and strong-wind frequency during each season are examined. The following conclusions from this study are delivered:

1. More numerous, yet weaker storms occur in the summer, while fewer, but stronger storms occur in the winter, which is consistent with the storm seasonal cycle for the pan-Arctic region (Zhang et al. 2004).
2. The storm intensity and high wind frequency correlated well from September to February, and June to August. Intense storms produce a significant impact upon the occurrence of strong surface winds.
3. Interplay between storm activity and Beaufort High helps to shape strong winds in the Chukchi-Beaufort Seas region. Strong wind could result from either the intense storms and/or strong Beaufort High. The large pressure gradient between the storm low pressures and Beaufort High is a favorable weather pattern for extremely strong winds.

Acknowledgements This work was supported by the Bureau of Ocean Energy Management of Department of the Interior under contract M06PC00018 and NSF Grants ARC-1023592 and PLR-1304684. Computing resources were provided by the Arctic Region Supercomputing Center at the University of Alaska Fairbanks.

References

Barker, D. M., et al. (2012). The weather research and forecasting model's community variational/ensemble data assimilation system: WRFDA. *Bulletin of the American Meteorological Society, 93*, 831–843.
Comiso, J. C. (2012). Large decadal decline of the Arctic multiyear ice cover. *Journal of Climate, 25*, 1176–1193.
Chen, F., & Dudhia, J. (2001). Coupling an advanced land-surface hydrology model with the PSU/NCAR MM5 modeling system. Part I: Model description and implementation. *Monthly Weather Review, 129*, 569–585.
Dee, D. P., Uppala, S. M., Simmons, A. J., Berrisford, P., Poli, P., & Kobayashi, S. (2011). The ERA-Interim reanalysis: Configuration and performance of the data assimilation system. *Quarterly Journal of the Royal Meteorological Society, 137*, 553–597. doi:10.1002/qj.828.
Grell, G. A., & Devenyi, D. (2002). A generalized approach to parameterizing convection combining ensemble and data assimilation techniques. *Geophysical Research Letters, 29*, 1693. doi:10.1029/2002GL015311.
Huang, X., et al. (2009). Four-dimensional variational data assimilation for WRF: Formulation and preliminary results. *Monthly Weather Review, 137*, 299–314.
Iacono, M. J., et al. (2008). Radiative forcing by long-lived greenhouse gases: Calculations with the AER radiative transfer models. *Journal of Geophysical Research, 113*, D13103. doi:10.1029/2008JD009944.
Janjic, Z. I. (2002). Nonsingular implementation of the Mellor–Yamada level 2.5 scheme in the NCEP Meso model. NCEP Office Note, No. 437, 61pp.
Liu, F., Krieger, J., & Zhang, J. (2013). Toward producing the Chukchi–Beaufort high-resolution atmospheric reanalysis (CBHAR) via the WRFDA data assimilation system. *Monthly Weather Review, 142*, 788–805.

Long, Z. X., & Perrie, W. (2012). Air-sea interactions during an Arctic storm. *Journal of Geophysical Research, 117*, D15103. doi:10.1029/2011JD016985.

Lynch, A. H., Curry, J. A., Brunner, R. D., & Maslanik, J. A. (2004). Toward an integrated assessment of the impacts of extreme wind events on barrow. *Alaska Bulletin of American Meteorological Society, 85*, 209–221.

Overland, J. E. (2009). Meteorology of the Beaufort Sea. *Journal of Geophysical Research, 114*, C00A07. doi:10.1029/2008JC004861.

Mellor, G. L., & Yamada, T. (1982). Development of a turbulence closure model for geophysical fluid problems. *Reviews of Geophysics and Space Physics, 20*, 851–875.

Morrison, H. C., Thompson, G., & Tatarskii, V. (2009). Impact of cloud microphysics on the development of trailing stratiform precipitation in a simulated squall line: Comparison of one- and two-moment schemes. *Monthly Weather Review, 137*, 991–1007.

Reed, M., et al. (1999). Oil spill modeling towards the close of the 20th century: Overview of the state of the art. *Spill Science & Technology Bulletin, 5*, 3–16.

Serreze, M. C., & Barrett, A. P. (2008). The summer cyclone maximum over the central Arctic Ocean. *Journal of Climate, 21*(5), 1048–1065.

Small, D., Atallah, E., & Gyakum, J. (2011). Wind regimes along the Beaufort sea coast favorable for strong wind events at Tuktoyaktuk. *Journal of Applied Meteorology and Climatology, 50*, 1291–1306.

Simmonds, I., Burke, C., & Keay, K. (2008). Arctic climate change as manifest in cyclone behavior. *Journal of Climate, 21*, 5777–5796.

Simmonds, I., & Keay, K. (2009). Extraordinary September Arctic sea ice reductions and their relationships with storm behavior over 1979–2008. *Geophysical Research Letters, 36*, L19715. doi:10.1029/2009GL039810.

Skamarock, W. C., Klemp, J. B., Dudhia, J., Gill, D. O., Barker, D. M., & Duda, M. G. (2008). A description of the advanced research WRF version 3. NCAR Technical Note NCAR/TN-475 +STR. doi:10.5065/D68S4MVH.

Stegall, S. T., & Zhang, J. (2012). Wind field climatology, changes, and extremes in the Chukchi–Beaufort seas and Alaska north slope during 1979–2009. *Journal of Climate, 25*, 8075–8089.

Zhang, X., & Zhang, J. (2001). Heat and freshwater budgets and their pathways in the Arctic Mediterranean. *Journal of Oceanography, 57*, 207–234.

Zhang, X., et al. (2004). Climatology and interannual variability of arctic cyclone activity: 1948–2002. *Journal of Climate, 17*(12), 2300–2317.

Zhang, X., et al. (2013). Beaufort and Chukchi seas mesoscale meteorology modeling study, final report. http://www.boem.gov/BOEM-2013-0119/. U.S. Dept. of the Interior, Bureau of Ocean Energy Management, AK. OCS Study BOEM 2013-0119. 204pp.

Designing a Remote In Situ Soil Moisture Sensor Network for Small Satellite Data Retrieval

Rawfin Zaman, William W. Edmonson, and Manoj K. Jha

Abstract This chapter introduces an application of small satellite for environmental monitoring using real-time data that focuses on measured soil moisture and temperature using in situ sensor network. Soil Moisture Active Passive Mission (SMAP) uses microwave radar and radiometer to sense surface soil moisture condition but gives coarser result than the in situ data. To overcome the limitation of accuracy in SMAP mission, the proposed architecture will provide sensor data via a Ground Monitoring Wireless Sensor Network (GM-WSN) where data is collected by small satellite(s) operating in Lower Earth Orbit (LEO). The satellite will store the data until it passes over a ground station whereby it is communicated back to earth. The motivation for developing satellite accessible in situ measurements is to retrieve information from remote areas like Lake Tana in Ethiopia, where human accessibility is difficult. Hence, those important and unattended places would be covered by the proposed system. Another key attribute of this architecture is that it addresses four (climate, carbon, weather, and water) out of six NASA's earth science strategic focus areas. The GM-WSN will consist of a sensor network that will measure soil moisture and temperature. The base station function is to fuse the data from the sensors, provide a time stamp, and format the data to be transmitted to satellite. It will also act as transceiver for ground to space communication using the amateur VHF/UHF radio band that has a maximum data rate of 9.6 kbps and will provide health maintenance and power management of network.

R. Zaman • W.W. Edmonson (✉)
Department of Electrical and Computer Engineering, North Carolina A&T State University, Greensboro, NC, USA
e-mail: rzaman@ncat.edu; wwedmons@ncat.edu

M.K. Jha
Department of Civil, Architectural and Environmental Engineering, North Carolina A&T State University, Greensboro, NC, USA
e-mail: mkjha@ncat.edu

© Springer International Publishing Switzerland 2016
G.A. Uzochukwu et al. (eds.), *Proceedings of the 2013 National Conference on Advances in Environmental Science and Technology*,
DOI 10.1007/978-3-319-19923-8_4

Introduction

In earth's atmosphere, hydrological cycle involves the continuous circulation of water including evaporation which causes the transfer of water from the surface of the earth to the atmosphere. Temperature is the main factor that causes evaporation. Soil is one of the principle sources from where water evaporation occurs. Desired amount of evaporation can be observed by knowing the water content in soil. Therefore, soil moisture and temperature are the two key parameters in hydrological cycle that need to be calibrated to ensure the understanding of climate change and global warming, develop improved flood prediction and drought monitoring capability, weather forecasting as well as the vegetation monitoring.

To obtain the measurement of soil moisture, various methods and models exist including feel method, gravitational method, tensiometer, electrical resistance blocks, neutron probe, phene cells, and time domain reflect meter methods. For the improved modeling of the soil moisture measurement, land surface models, all water and energy balance models, general circulation models, weather prediction models, and ecosystem process simulation models are prevailing (Yang 2004). Depending on the particular application area for high-resolution, continuous extensive data sets and to achieve below-ground observations, ground-truth data sets in situ sensor-based wireless network is now extensively required. To cover a distributed area, remote wireless sensor network (WSN) is an appropriate choice because it plays a vital role when there is a need of detailed monitoring of remote and inaccessible locations. It is made up of cluster of sensor nodes that has a capability to reconfigure itself in an ad hoc fashion that is used for various earth observation applications.

Numerous studies and missions are ongoing for measuring soil moisture, soil temperature using satellite footprint to cover large, unattended places. By the NASA's SMAP mission (The Soil Moisture Active and Passive Mission 2008), targeted to launched Jan 31, 2015, uses active radar and a passive radiometer that operates at L-Band frequency will make global measurements of soil moisture and its freeze/thaw-state (Spencer et al. 2011). The SMAP mission, in situ calibration is needed in order to compare the data extracted from satellite footprint images and the real-time land data (Cano et al. 2007). Soil moisture is highly changeable with the dielectric constant at low microwave frequency that leads to a coarse result than the actual situation. The Soil Moisture and Ocean Salinity (SMOS) (Ocean Salinity 2004) mission's objective is to provide global maps of soil moisture (SM) and sea surface salinity (SSS). This system retrieves soil moisture using brightness temperature observations. It consists of a satellite in LEO operating at L-band that acquires passive radiometric information of earth using ground maps of brightness temperature at different polarizations for all land and ocean areas imaged (Barre et al. 2008). The soil emissivity, "epsilon" depends upon the moisture content at a particular microwave frequency. It's been observed that there is a large dielectric contrast between dry soil and water. For example, at L-band, in particular, the sensitivity of dielectric permittivity to soil moisture is very high, whereas

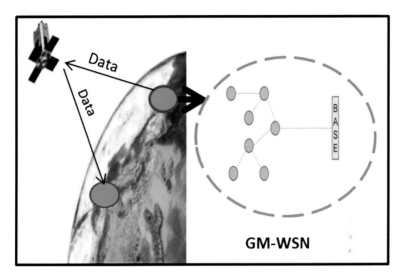

Fig. 1 Small satellite data retrieval

sensitivity to surface roughness is minimal (SMOS 2000). To eliminate these limitations, the proposed mission envisions, in situ WSN system using small satellite data retrieval capability for environmental monitoring as shown in Fig. 1.

Background study of the system. The goal of the system design is to assist environmental monitoring by collecting real-time data on soil moisture and soil temperature. Soil moisture data helps to enhance the agriculture productivity, augment the weather and climate forecasting ability, enhance flood prediction and drought monitoring aptitude, early warning of diseases, and comprehend the process of water, energy, and carbon cycle. Our proposed system will help agriculture productivity by providing irrigation scheduling, crop yielding by stipulating the information on water availability in soil. Numerical weather prediction and climate model could improve forecasting severe rainfall by using consistent soil moisture data considerably in continental region. Flooding and landslides can occur with the presence of high soil moisture. Precise soil moisture reduces the risk of flash flood and river flow forecasts. Indication of drought is noticeable when there are consistent low soil moisture values. More frequent, accurate and spatially complete soil moisture measurements are critical to drought monitoring and early warning. Numerous kinds of diseases are found around Lake Tana in Ethiopia, South Asia, and Lake Victoria due to monsoonal weather. The proposed system will help for the early warning of these monsoonal diseases (Entekhabi et al. 2010).

Small satellites have opened a new era in space technology by decreasing space mission costs, without greatly reducing the performance. Prior to transmitting the soil moisture data to the small satellite, a GM-WSN is proposed that gathers in situ data. WSN has the capability of fast deployment, can accommodate new devices at any time, is flexible to go through physical partitions, and can be accessed through a centralized monitor. The soil moisture in situ data from GM-WSN will be transmitted to the small satellite and the satellite will work in a store and forward mode. It will downlink the stored data when it will be overhead to another Ground Observing Station (GOS).

Design issue. Our system vision is to cover distributed environmental monitoring, ensuring reliability, longer lifetime, ease of deployment and maintenance, the ability to support low cost and relatively long-range communication, and also reduction in cost. In SMAP mission, the L-band radiometer measurement provides an estimation of soil moisture 5 cm depth from the earth's surface (Moran et al. 2009). Our system's goal is to improve the penetration depth to measure the soil profiles. Deeper layer soil moisture observations give better soil moisture results. The capacitance probe soil moisture sensor in the GM-WSN system helps to achieve the improvement, which has approximately 10 cm penetration depth. The SMAP mission often gives unlike results between the footprint images taken by the satellite and the in situ ground-truth assessment because soil moisture is highly heterogeneous in ground due to varying climate conditions, soil characteristics, and vegetation coverage (Moghaddam et al. 2010). SMOS mission aims to monitor over the ground and sea surface with enough resolution that need to be validated with ground measurement (Cano et al. 2007). Our proposed system architecture aims to overcome the limitation in both the SMAP and SMOS mission by acquiring the real-time land data with the help of ground segment and direct that to the small satellite. Numerous sensors connect to the assigned stimulator at different required depth and form a sensor node. The stimulator has the capability to communicate wirelessly with the central coordinator. The coordinator schedules and transmits the scheduling command to each node, and receives the sensor readings from the sensor node thus make a WSN. The coordinator then sends the in situ sensor data to the processor in the base station, after completing the required processing; processed data is transmitted to the small satellite.

System architecture. The system architecture shown in Fig. 2 is divided into two sections, namely, ground segment and space segment. The workflow in the ground segment is allocated into GM-WSN and base station for collecting surface-to-depth soil profiles and performs the data processing before transmitting to the satellite.

Ground segment. To achieve the aim of accumulating the surface-to-depth soil profiles at distributed locations, our system requires a network containing soil moisture, temperature sensors, and wireless modules that has a capability to work as wireless RF transceiver referred to as sensor node. The sensor nodes are the RF module which stimulate and regulate the sensor probes and communicate with the base station to send the collected data periodically to the base station. A detailed description about the network topology, sensor nodes, sensor, and the base station for the proposed system are described below:

Ground monitoring wireless sensor network. Automated multisensor system is apposite for the ground system as it is compatible for remote places. Automated multisensor-node system consists of a set of electronics that includes sensors, microprocessor, RF module, and GPS.

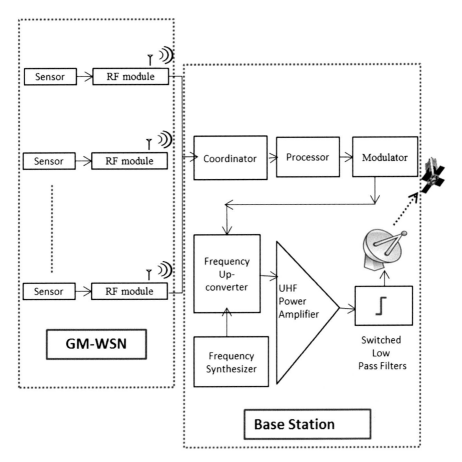

Fig. 2 System architecture

Network topology. After surveying on the current wireless technologies including Wi-Fi, Wi-max, Bluetooth, etc., it is observed that currently ZigBee (Zigbee 2002) technology is the only standard-based technology in the market that targets low data rate, low power consumption, low cost, and wireless networking protocol. Zigbee covers up to 90 m to 1 mile line of sight areas for data transfer which is more feasible for WSN communication compared to other IEEE 802 standards. It ensures our wide range communication to cover distributed observation. ZigBee is built on top of the IEEE 802.15.4 standard which defines the Medium Access Control (MAC) and physical layers, operating in an unlicensed band of 2.4 GHz with a data transfer rate of 250 kbps. Zigbee aims to provide network, security, and application layers. The ZigBee standard has the capacity to address up to 65,535 nodes in a single network and all the network designs need to follow the following three types of logic devices:

(a) *Coordinator*: All ZigBee networks must have one coordinator. The tasks of the Coordinator at the network layer are

- Starts the network
- Allows other devices to connect to it

(b) *End device*: End Devices are always located at the extremities of a network

- The main tasks of an End Device at the network level are sending and receiving messages
- End Devices cannot relay
- It sleeps in order to conserve power

(c) *Router*: The main tasks of a Router are

- Data routing, relay messages from the coordinator, end devices, and other routers

It follows mesh, tree, cluster configuration, and also star-mesh hybrid topology. To achieve robustness, scalability and to reduce complexity our proposed system supports cluster-tree topology (Fig. 3). The transmission range of each node is 90 m and in our system model the separation distance between the nodes is set to be 90 m. Therefore, in order to cover the whole cluster area at least six nodes are required. For a cluster with n number of nodes and the same number of interconnection with other cluster, we found that entire number of nodes (N) in the system is calculated as follows:

$$N = (n - ic) \times C + R \qquad (1)$$

where n represents the number of nodes in each cluster, ic represents the number of interconnection between clusters, C represents the number of cluster, and R represents number of routers.

Adding more clusters in tree topology will increase the coverage area, with radius 90 m the coverage area by one cluster is 25,446 m^2. Table 1 illustrates the area covered by the number of clusters:

RF module. For building the communication between the network devices appropriate sensor node needs to be chosen. After choosing the Zigbee technology to generate the sensor node, XBEE ZB S2 module has been proposed. XBEE ZB S2 is

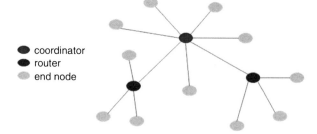

Fig. 3 Zigbee cluster-tree topology (Moran et al. 2009)

Table 1 Coverage area with the number of cluster

Router	Number of data point	Coverage area (m^2)
1	5	25,446
2	10	50,892
3	15	76,338

a Zigbee-complaint WSN device developed by Digi International. More details on the Zigbee enabled device XBEE ZB S2 can be found in (XBee PRO ZB 1996). Different firmware versions of XBEE PRO ZB is compatible to act as one of the three logic device types in a Zigbee network and we will use it in this way, the mother node of our system is the coordinator and connects to the base station computer through USB port. The base station computer is programmed with X-CTU software to ensure information flows from base station to the network and back. The sensors are connected to the end device through its serial port without using any additional translator. For the workload of this device, it requires more energy and to ensure longer lifetime rechargeable power supply is proposed in our system for each node. End device and router has the same configuration in our system, it establishes the communication between the end devices and the coordinator.

Sensor. The 5TM sensor from Decagon device is chosen for the proposed system because it is an appropriate in situ sensor compatible for measuring both soil moisture and temperature simultaneously thus replacing the use of two separate sensors. Decagon device capacitance probe sensors are widely used in the field of WSN applications and research (Barre et al. 2008; Moghaddam et al. 2010). It gives up to 15 % accuracy for the volumetric water content and ± 1 °C for the temperature ± 1 "epsilon" for the dielectric constant. The sensor measures the dielectric constant using electromagnetic field of the surrounding soil medium to determine soil moisture. The sensor generates 70 MHz oscillating wave to the prongs that charges according to the dielectric of the material. The stored charge is proportional to soil dielectric and soil volumetric water content (Li et al. 2011). The automated multisensor consists of a microprocessor that measures the charge and outputs a value of dielectric permittivity from the sensor. A thermostat which is mounted in the sensor next to one of the prongs is used to take temperature readings of the prong surface. It gives the output result in Kelvin and at 0.1 °C resolutions. For soil moisture measurement, the resolution is 0.1 "epsilon" for dielectric constant 1–20 and 0.75 "epsilon" for dielectric constant 20–80. The GM-WSN is flexible due to open and modular architecture; more sensors can be added to measure different parameters as electrical conductivity, pressure, etc. according to the requirement.

This sensor gives the digital output in ASCII code. Table 2 illustrates key characteristics of 5TM Decagon sensor.

Sensor communication. RF module excites the sensor by applying an excitation voltage for 120 ms which gives three measurement results and is transmitted to the RF module. The output is eight data bits ASCII characters with one stop bit and no parity. The voltage level is 0–3.7 V and logic level is TTL (active low). The ASCII stream contains three numbers separated by spaces where the first number is raw

Table 2 Specification of 5TM sensor

Specification	5TM sensor
Resolution (soil moisture)	0.1 (for dielectric constant 1–20) 0.75 (for dielectric constant 20–80)
Resolution (soil temperature)	0.1° C (for −40 to 50° C)
Dielectric measurement frequency	70 MHz
Measurement time	150 ms (milliseconds)
Power requirements	3.6–15 V-DC, 0.3 mA quiescent, 10 mA during 150 ms measurement
Output RS232	8 bits ASCII character
Operating temperature	Operating temperature −40 to 50° C

dielectric output and the third one is raw temperature, second value is zero which is ignorable. The dielectric constant in air is 1 and water is 80. The sensor gives the output in the range of 0–4094 which corresponds to the dielectric permittivity value of 0.00–81.88. When the output gives 4095, it indicates that the sensor is not working as expected (Li et al. 2011). To convert the raw dielectric permittivity value into actual dielectric permittivity value, the following equation (2) is given:

$$\text{Dielectric permitivity} = \text{Dielectric permitivity(raw)}/50 \qquad (2)$$

This sensor connects and communicates with the Xbee RF module which has been chosen for this work through serial TTL UART port.

Power supply. In the GM-WSN, end device connects sensors through its serial port to collect sensor data and requires more power for its heavy workload and for its long radio range communication. By the manufacturer, each node can run for 1 year with a regular battery with 1-year lifetime. Therefore, rechargeable power supply (RPS-1) is proposed to ensure longer lifetime. RPS-1 is an autonomous battery power source, which can regulate output voltage from 3.5 V-DC to 22 V-DC (5TM Operator's Manual 2010). The functional block diagram for the RPS-1 is shown in Fig. 4.

It has two user-programmable output voltage supplies and the supply voltage is regulated as required. Voltage regulator 1 and 2 is connected to the XBEE module and the sensors, respectively, to ensure the power supply.

Data flow. Each sensor gives the three outputs within 120 ms time and the output is 1200 baud asynchronous with eight data bits ASCII characters. The ASCII stream contains three numbers separated by spaces where first number represents the dielectric output and the third number is temperature and the second number is zero to separate the two values. The sensor gathers 249.6 kbits of data within 20.8 min, with this data rate, hourly output of the sensor is 90 kbyte. The sensor transmits the data to the RF module which communicates with other RF module with a data rate of 250 kbits/s. The communication between the RF modules is stated below in Fig. 5.

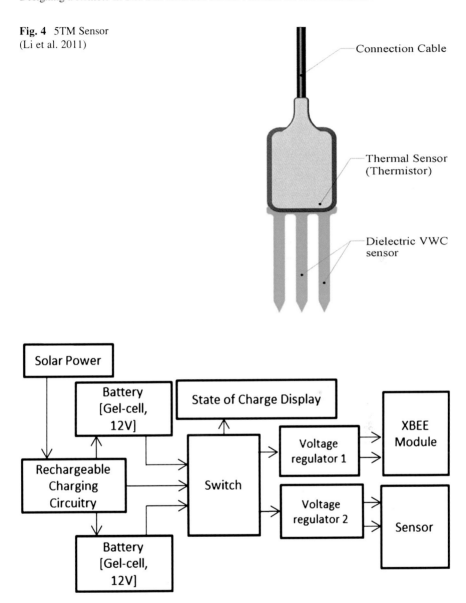

Fig. 4 5TM Sensor (Li et al. 2011)

Fig. 5 Functional block diagram for power supply

Uplink data can be calculated by (3).

$$\text{Uplink data} = N^*(\text{Data}_N + \text{Overhead}) \qquad (3)$$

where N symbolize number of nodes.

Data$_N$ symbolize the data rate per node

Overhead = PHY Layer Overhead + MAC Layer Overhead
= (Preamble Sequence + Start of Frame Delimiter + Frame Length)
+ (Frame Control + Data Sequence Number + Address Information
+ Frame Check Sequence)
= (4 + 1 + 1) bytes + (4 + 1 + (4 − 20) + 2) bytes
= 15 to 31 Bytes

In summary, the total overhead for a single packet can range from 15 to 31 bytes.

Base station. Earth station design is important as is the design of satellite space craft. Earth station can operate either in receiving or transmission mode or in both receiving and transmission mode. In the proposed system, the ground segment drives in transmission mode. The collected data from the ground monitoring WSN will be transmitted to the base station with the help of coordinator node. The coordinator node (XBEE Pro) has the serial connection with the processor. In the processor, the data processing, encryption, and decryption take place. The data from the base station (BS) will link to the satellite when the satellite will be in the LOS of the BS, transmit the acquired sensor data to the satellite. The data is then transmitted to a satellite the goes overhead to another GOS; the satellite transmits the data to the GOS and thus driving it in receiving mode (Fig. 6).

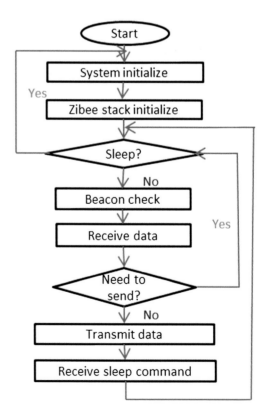

Fig. 6 Data flow chart

Conclusion and Future Work

The system introduces a WSN system with the small satellite data retrieval capacity. It is fundamentally designed for low cost, easy deployment, and essential system model. The ground segment considered here aims to use in situ sensors with high resolution to gather soil moisture and temperature data as well as to cover small satellite application. The soil moisture sensors are highly accurate and ensure improved sensing depth. Zigbee technology ensures the high data rate with low cost, low power consumption. For the proposed system, we will use COTS (Commercials off the Shelf) design for space communication linkage. As rechargeable power supply has been proposed, the system is sustainable for longer period of time. By using cluster-tree topology, it ensures larger coverage, more reliable and scalable. The future work is to use cluster-tree topology that has an improved coverage area by adding more clusters in the system that requires multi-hop communication. Therefore, intelligence in the node needs to be added to ensure the multi-hop, thus the node will operate as agent. Another upcoming issue is that, we can operate the node as a device, so that it will collect the soil moisture data only at the certain required level.

References

Barre, H. M. J., Duesmann, B., & Kerr, Y. H. (2008). SMOS: The mission and the system. Paper published in IEEE transactions on geoscience and remote sensing, Vol. 46.

Cano, A., et al. (2007). Wireless sensor network for soil moisture applications. *Paper Published in International Conference on Sensor Technologies and Applications*, Valencia, 14–20 October 2007.

Entekhabi, D., et al. (2010). The Soil Moisture Active Passive (SMAP) mission. This paper describes an instrument designed to distinguish frozen from thawed land surfaces from an Earth satellite by bouncing signals back to Earth from deployable mesh antennas. *Published in Proceedings of the IEEE*, Vol. 98. pp. 704–716.

Li, K., Jia, H., & Liu, M. (2011). ZigBee wireless sensor network using physics-based optimally sampling for soil moisture measurement. Third international conference on communications and mobile computing (CMC), Qingdao, 18–20 April 2011.

Moghaddam, M., Entekhabi, D., Goykhman, Y., et al. (2010). A wireless soil moisture smart sensor web using physics-based optimal control: concept and initial demonstrations. *IEEE Journal of Selected Topics in Applied Earth Observations and Remote Sensing*, doi:10.1109/JSTARS.2010.2052918:522–535.

Moran, S., et al. (2009). Report of the 1st SMAP applications workshop. Silver Spring, MD, September 9–10, 2009.

5TM Operator's Manual. (2010). Decagon Devices, Pullman, WA. Retrieved May 25, 2013, from http://www.decagon.com/education/5tm-manual/.

RPS-1 Portable Power Supply (2012). Retrieved May 28, 2013, from www.logicbeach.com/pdf/RPS-PowerSupply.pdf.

Soil Moisture and Ocean Salinity (2004). Retrieved April 23, 2013, from http://www.esa.int/Our_Activities/Observing_the_Earth/SMOS.

SMOS (Soil Moisture and Ocean Salinity) mission. (2000–2013). Retrieved May 5, 2013, from https://directory.eoportal.org/web/eoportal/satellite-missions/s/smos.
Spencer, M., et al. (2011). The planned Soil Moisture Active Passive (SMAP) mission L-band radar/radiometer instrument. Paper published in 2011 I.E. international geoscience and remote sensing symposium (IGARSS), Vancouver, BC, 24–29 July 2011.
The Soil Moisture Active and Passive Mission. (2008). Retrieved April 21, 2013, from http://smap.jpl.nasa.gov/.
XBee PRO ZB module, Digi International (1996–2013). Retrieved May 24, 2013, from http://www.digi.com/products/wireless-wired-embedded-solutions/zigbee-rf-modules/zigbee-mesh-module/xbee-zb-module.
Yang, Z.-L. (2004). Modeling land surface processes in short term weather and climate studies. In X. Zhu (Ed.), *Observation, theory and modeling of atmospheric variability* (World scientific series on meteorology of East Asia, pp. 288–313). Hackensack, NJ: World Scientific.
Zigbee. (2002). Retrieved May 25, 2013, from http://www.zigbee.org/.

Alaskan Regional Climate Changes in Dynamically Downscaled CMIP5 Simulations

Jing Zhang, Jeremy Krieger, Uma Bhatt, Chuhan Lu, and Xiangdong Zhang

Abstract Global models are the most widely used tools for understanding and assessing climatic variability and changes. However, coarse-resolution limits their capability to capture detailed finer-scale meteorological features, including heterogeneous spatial distributions and high-frequency temporal variability. In this study, the mesoscale Weather Research and Forecasting (WRF) model is used to dynamically downscale a CMIP5 global model simulation (CCSM MOAR output) for a portion of the Arctic marginal zone, encompassing Alaska and surrounding areas, with the aim to improve understanding, representation, and future projection of high-resolution climate changes in the area. Dynamic downscaling of the twentieth century simulation was conducted for the period 1991–2005 and validated against in situ observations archived by the NCDC. Downscaled results generally capture observed conditions well. However, cold biases exist across most of the study area, except for a weak warm bias along the western and northern Alaskan coasts. In addition, downscaled winds are stronger than observations and precipitation is

J. Zhang (✉)
Department of Physics, North Carolina A&T State University,
Greensboro, NC 27411, USA

Department of Energy and Environmental Systems, North Carolina A&T State University,
Greensboro, NC 27411, USA
e-mail: jzhang1@ncat.edu

J. Krieger
Arctic Region Supercomputing Center, University of Alaska Fairbanks,
Fairbanks, AK 99775, USA

U. Bhatt
Geophysical Institute, University of Alaska Fairbanks, Fairbanks, AK 99775, USA

C. Lu
International Arctic Research Center, University of Alaska Fairbanks,
Fairbanks, AK 99775, USA

Nanjing University of Information Science and Technology, Nanjing, China

X. Zhang
International Arctic Research Center, University of Alaska Fairbanks,
Fairbanks, AK 99775, USA

© Springer International Publishing Switzerland 2016
G.A. Uzochukwu et al. (eds.), *Proceedings of the 2013 National Conference on Advances in Environmental Science and Technology*,
DOI 10.1007/978-3-319-19923-8_5

overestimated along the Alaskan panhandle. The biases in the downscaled temperature, wind speed, and precipitation are correctable. The downscaled temperature bias exhibits strong seasonality, with a warm bias in the cold months and a cold bias in the warm months, particularly along the western and northern Alaskan coasts. Seasonality in the wind speed and precipitation biases, however, is relatively small. Under the RCP6 scenario, downscaled regional climate over Alaska and the surrounding areas demonstrate a significant warming trend over the entire study area during the twenty-first century, with the strongest warming occurring over the Arctic Ocean. Precipitation is also projected to increase along Alaska's coastal areas and over the Arctic Ocean. Interior Alaska, on the other hand, becomes drier in the future climate scenario.

Introduction

Alaska and its surroundings are undergoing significant environmental changes, including a large increase in surface air temperature (Whitfield 2003), a rapid decline in glacier mass balance (Hock et al. 2009; Gardner et al. 2013), continued thawing of the permafrost (Osterkamp 2003), and an accelerated decline in sea ice coverage in the Chukchi and Beaufort Seas (Comiso et al. 2008; Stroeve et al. 2012), along with increased surface wind speeds as the sea ice retreats (Stegall and Zhang 2012). These changes motivate a number of urgent questions: How will these changes evolve during the next half-century under the influence of a continued increase in greenhouse gas emissions, and how can we best prepare in order to adapt to these coming changes? To address these concerns, accurate information about climate change in Alaska is needed.

While general circulation models (GCMs) allow for global-scale climate simulation and projection under a range of scenarios, the coarse spatial (100 s of km) and temporal resolution of GCM output makes it difficult to accurately assess the local and regional impacts of climate change. This is particularly true for regions with complex topography (Leung et al. 2003). Alaska and its surrounding areas represent prominent geographical features, with seasonal ice coverage over the ocean and sharply varying terrain on land. A comparison of the annual mean precipitation over Alaska as resolved by the National Centers for Environmental Prediction/National Center for Atmospheric Research (NCEP/NCAR) reanalysis (~2.5° grid spacing) (Kalnay et al. 1996) and mesoscale model simulations (10–30 km grid spacing) demonstrates that finer-scale structures associated with terrain effects can only be captured in high-resolution simulations (Zhang et al. 2007). Thus, there is a growing need to develop downscaling methodologies (e.g., empirical or dynamical) to quantitatively obtain regional- and local-scale climate change information from coarse-resolution GCM output.

The dynamical regional climate downscaling technique is a commonly used approach and has been applied in order to better represent and understand local weather systems and their associated impacts (Bengtsson et al. 1996; Lynch et al. 1998; Zhang et al. 2007; Giorgi et al. 2009; Mearns et al. 2009). The Coordinated Regional Climate Downscaling Experiment (CORDEX) (Giorgi et al. 2009) and the North American Regional Climate Change Assessment Program (NARCCAP) (Mearns et al. 2009) are examples of such ongoing efforts. The

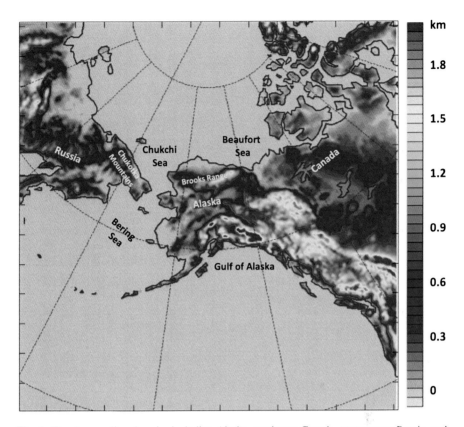

Fig. 1 The downscaling domain, including Alaska, northwest Canada, easternmost Russia, and the surrounding ocean including the Beaufort, Chukchi, and Bering Seas

downscaled domains included in CORDEX are primarily used to represent continental regions around the world and NARCCAP encompasses most of North America. Unfortunately, a downscaling domain over the entirety of Alaska is not included in either the CORDEX or NARCCAP efforts, thus necessitating further downscaling activities focused specifically on Alaska. These focused efforts are needed to safeguard future Alaskan economic development, environmental safety, and security.

The Coupled Model Intercomparison Project 5th phase (CMIP5) global climate simulations and projections represent the latest coordinated effort climate modeling around the world and provide important resources for the Intergovernmental Panel on Climate Change (IPCC) 5th Assessment Report (AR5). In this study, we employ the dynamical downscaling method to downscale CMIP5 simulations and projections for the Alaska region (Fig. 1) using a physically optimized version of the Weather Research and Forecasting (WRF) modeling system (Skamarock et al. 2008). The remainder of this chapter is structured as follows: Section "Data and Model" provides a brief description of the GCM data and downscaling methodology used in this study; Section "Dynamical Downscaling of the Twentieth century CCSM4 Simulation" describes the calibration of the twentieth century downscaled regional climate throughout Alaska with in situ station observations; the downscaled future climate features over Alaska are summarized in section "Downscaled Future Climate Changes Under RCP6 Scenario over Alaska"; and a summary of this study is given in section "Summary."

Data and Model

The AR5 twentieth and twenty-first century simulations and projections produced with the state-of-the-art GCM NCAR Community Climate System Model 4.0 (CCSM4) are selected to conduct regional climate downscaling over Alaska and its surroundings (Gent et al. 2011). The CCSM4 results are used to provide the initial conditions and boundary forcing for the WRF downscaling simulations based on our previous experience using MM5 (Mesoscale Model version 5, the predecessor to WRF) for regional downscaling of CCSM3 (previous version of CCSM4) simulations (Zhang et al. 2007). The downscaling domain (Fig. 1) employed by the WRF model has a grid spacing of 20 km, covering Alaska along with parts of northwestern Canada, northeastern Russia, and the surrounding ocean. The twentieth century all-forcing CCSM4 simulations from the MOAR (mother of all runs) ensemble member are downscaled to the modeling domain for the period 1991–2005. The downscaled present-day climate over Alaska is verified against in situ observations archived by the National Climatic Data Center (NCDC) to calibrate the downscaling performance. Around 505 stations used to calibrate temperature and wind speed and 406 stations used to calibrate precipitation are acquired from the NCDC across the study domain over the period 1991–2005.

In order to assess regional climate changes and impacts in Alaska, the CCSM4 future climate projections under the Representative Concentration Pathway 6 (RCP6) mitigation scenario and from the same MOAR ensemble member used for the historical runs are selected for downscaling. The selected RCP6 scenario is a middle-of-road estimate of future emissions, representing a relatively probable outcome for future climate conditions. In RCP6, the total radiative forcing increases to a maximum of about 6 W/m^2 in the year 2100, after which it is stabilized without overshoot. Likewise, the atmospheric CO_2 concentration rises to a value of around 750 ppm in 2100 before stabilizing. Considering that a continuous meteorological forcing is needed for projecting future changes relative to the current climate, such as changes to glaciers and permafrost, a downscaled regional climate for the entire twenty-first century is needed. To this end, the entire twenty-first-century (2005–2100) CCSM4 projection under the RCP6 scenario is also downscaled over Alaska. All the CCSM4 results are acquired from the Earth System Grid Federation (ESGF).

A physically optimized configuration of WRF model physical parameterizations for Alaska (Zhang et al. 2013) is adopted to conduct the downscaling simulations in this study. In addition, in performing the downscaling simulations with WRF, the

Table 1 WRF model configuration for downscaling simulations

	Options	Configuration
Physics	Microphysics	Morrison 2-moment (Morrison et al. 2009)
	Longwave radiation	Rapid Radiative Transfer Model for GCMs (RRTMG) (Iacono et al. 2008)
	Shortwave radiation	RRTMG (Iacono et al. 2008)
	Cumulus	Grell 3D ensemble cumulus (Grell and Devenyi 2002)
	Planetary boundary layer	Mellor–Yamada–Janjic (Eta) (Mellor and Yamada 1982; Janjic 2002)
	Surface layer	Monin–Obukhov (Janjic Eta) (Janjic 1994, 1996, 2002)
	Land-surface model	Noah land-surface model (Chen and Dudhia 2001) coupled with a thermodynamic sea ice model (Zhang and Zhang 2001)
Grid	Horizontal grid spacing	20 km
	Vertical levels	49 levels with top at 10 hPa
Nudging	Spectral nudging	Wave number 3 for all variables at all levels

use of a nudging technique is essential to ensure that the model does not deviate significantly from the GCM forcing. Following Zhang et al. (2013), we apply spectral nudging with a wave number of 3 and nudge all variables at all vertical levels in the downscaling simulations. The detailed WRF model configuration for this study is summarized in Table 1. It should be noted that a thermodynamic sea ice model (Zhang and Zhang 2001) is coupled with the Noah land-surface model within WRF in order to accurately model the thermal conditions over sea ice.

Dynamical Downscaling of the Twentieth Century CCSM4 Simulation

The downscaled twentieth century regional climate (as represented by surface temperature, wind speed, and precipitation) generated by WRF exhibits clear seasonal variability (Fig. 2). Except for summer, seasonal surface temperatures over Alaska are generally colder than over the North Pacific Ocean and warmer than over the Arctic Ocean (Fig. 2a–d). Summer surface temperatures, on the other hand, particularly over the lowlands of Alaska, are much warmer over land. The annual amplitudes of seasonal mean surface temperature variation in Alaska are around 30 K along the north coast, 10–20 K along the west coast, 20–40 K in the Interior, and less than 10 K over the Gulf of Alaska. Mountain impacts on surface temperatures are readily apparent, with colder surface temperatures over mountains seen throughout the year.

Surface winds across the entire downscaling domain also show a seasonal cycle, with stronger winds in winter and fall, weaker winds in spring, and relatively calm winds in summer (Fig. 2e–h). Strong winds due to intense storm activity during

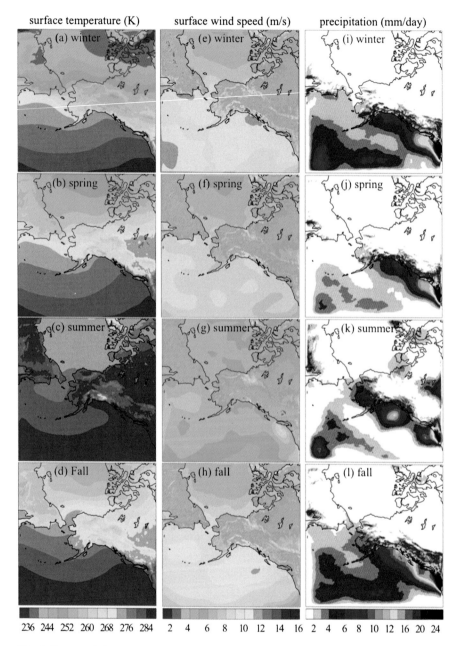

Fig. 2 Downscaled seasonal mean surface temperature (K, (**a**)–(**d**)), surface wind speed (m/s, (**e**)–(**h**)), and precipitation (mm/day, (**i**)–(**l**)) during 1991–2005

winter and fall are observed over the North Pacific Ocean. Surface winds over the Bering, Chukchi, and Beaufort Seas are enhanced as the sea ice retreats northward. The winds over the Chukchi and Beaufort Seas are strongest during fall when minimum sea ice coverage is present. Overall, winds over land are weaker than over the ocean, including both open and ice-covered waters, except for mountain areas, which experience consistently stronger winds. Seasonality of surface winds over land is relatively small.

Most precipitation occurs in the southern part of downscaling domain, with strong precipitation centers along the southeast coast of Alaska and west coast of Canada (Fig. 2i–l). Mountain lifting effects result in strong coastal precipitation, with many detailed features captured by the high-resolution model. Over Alaska, precipitation primarily occurs along the south and southwest coasts and southeast panhandle. Over the southwest coastal areas, precipitation primarily occurs during summer and fall. Influenced by the storm track over the North Pacific Ocean, strong precipitation occurs during fall and winter.

The WRF-downscaled CCSM4 twentieth century simulations are also compared with station observations to calibrate the performance and estimate biases in the downscaled simulations. This comparison provides a measure with which to estimate errors due to model biases in the WRF downscaling of future CCSM4 projections. The downscaled parameters at 20-km grid spacing are interpolated to the locations of weather stations with the Cressman interpolation technique, taking the surface type (land vs. water) and elevation information into account. When interpolating, elevation limits were set to exclude grid points with elevations different from the station elevation by more than 100 m, as interpolated values show much greater sensitivity to elevation than to horizontal distance. Comparisons between the interpolated WRF-downscaled results and station observations, as averaged over the entire simulation period 1991–2005, show that persistent biases exist in the downscaled climate (Fig. 3). Relatively large cold biases exist in southwest Alaska and the Alaska panhandle, while a slightly warm bias occurs along the west and north Alaska coasts. The downscaled winds are overall stronger than observed values for the entire downscaling domain. While there are far fewer observed precipitation data available in the study domain, the precipitation bias demonstrates that precipitation is overestimated along the Alaska panhandle and underestimated along the south Alaska coast.

Biases in the downscaled surface temperatures exhibit strong seasonality (Fig. 4). Aside from cold biases in the Alaska panhandle, which occur throughout the year, warm biases along the west and north Alaska coasts occur only during the cold months, while cold biases in southwest Alaska are larger during the warm months. Overall, the downscaled surface temperatures in the study domain are colder than observations, except along the west and north Alaska coasts where warm biases occur during the cold months. The seasonality in wind speed and precipitation biases, on the other hand, is relatively small (not shown).

Considering that the biases in downscaled climate parameters, particularly surface temperature, demonstrate a seasonal variation, and following the method

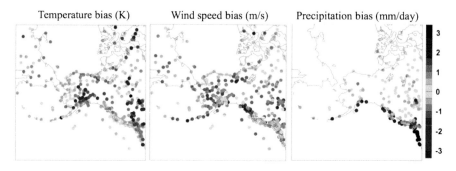

Fig. 3 Annual mean biases (downscaling minus observations) of the downscaled surface temperature (K), wind speed (m/s), and precipitation (mm/day), as verified against in situ observations for the period of 1991–2005

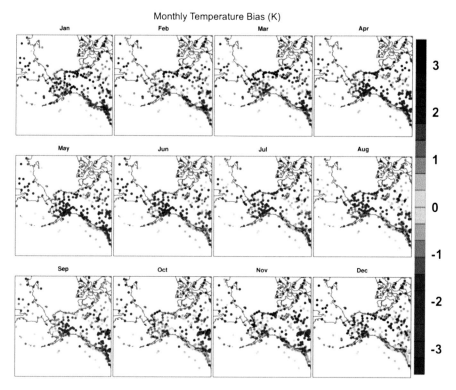

Fig. 4 Monthly mean biases of the downscaled surface temperature (K), as verified against in situ observations for the period 1991–2005

used in Zhang et al. (2007), mean monthly biases can be calculated by computing the differences between the monthly averages of station observations and the WRF-downscaled output. The WRF-downscaled temperature, wind speed, and precipitation are then corrected as follows:

$$T^c_{(d,m,y)} = T^u_{(d,m,y)} + dT_{(m)}$$
$$W^c_{(d,m,y)} = W^u_{(d,m,y)} + dW_{(m)}$$
$$P^c_{(d,m,y)} = P^u_{(d,m,y)} + \frac{P^u_{(d,m,y)}}{P^u_{(m,y)}} \times dP_{(m)} \quad (1)$$

where subscripts d, m, and y refer to the dth day of the mth month in the yth year; superscripts c and u represent corrected and uncorrected variables, respectively; $T^{u(c)}_{(d,m,y)}$ is daily mean temperature; $W^{u(c)}_{(d,m,y)}$ is daily mean surface wind speed; $P^{u(c)}_{(d,m,y)}$ is daily precipitation; $P^u_{(m,y)}$ is monthly precipitation; and $dT_{(m)}$, $dW_{(m)}$, and $dP_{(m)}$ are averaged monthly biases of temperature, wind speed, and precipitation for the mth month, respectively. Following Zhang et al. (2007), the mean monthly precipitation correction $dP_{(m)}$ was distributed over the daily precipitation $P^u_{(d,m,y)}$ with the weighting function $P^u_{(d,m,y)}/P^u_{(m,y)}$, the ratio of daily to monthly precipitation, to ensure that a larger precipitation correction is applied on days with large precipitation amounts than on those with little rain.

As an example, the corrected and uncorrected 15-year (1991–2005) average daily temperatures at Anchorage are compared with observations (Fig. 5). Compared with the observed temperatures (black dots), the uncorrected temperatures

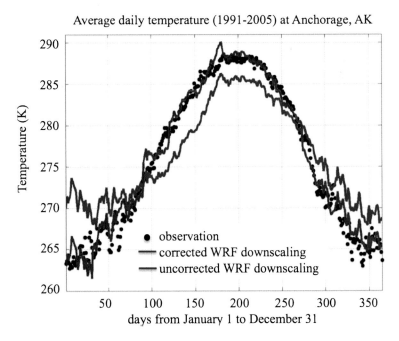

Fig. 5 Comparisons among observations (*black dots*), WRF downscaling (*blue curve*), and corrected WRF downscaling (*red curve*) for the daily mean temperature at Anchorage, AK

(blue curve) tend to be warmer during cold months and, conversely, colder during warm months. These seasonal biases are largely removed by applying (1) to the WRF-downscaled output. Corrected temperatures show reasonable agreement with the observations. The biases in the WRF downscaling are thus correctable.

Downscaled Future Climate Changes Under RCP6 Scenario over Alaska

The WRF-downscaled future climate forced by the CCSM4 twenty-first century RCP6 scenario projections is corrected with the biases calculated from the twentieth century downscaling (1991–2005). That is, the biases and bias corrections identified and applied for the past simulations are assumed to be time-independent and applied to the future downscaling results. Figure 6 depicts the

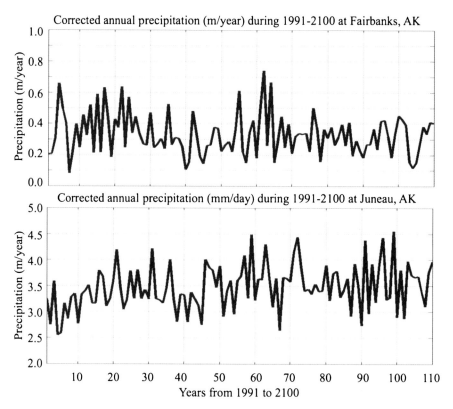

Fig. 6 Bias-corrected WRF downscaling of annual precipitation at Fairbanks (*top*) and Juneau (*bottom*), AK for the entire downscaling period of 1991–2100

evolution of bias-corrected WRF-downscaled annual precipitation for the entire downscaling period of 1991–2100 at Fairbanks and Juneau, Alaska. Changes at Fairbanks are used to represent the future scenario for the interior of Alaska, while Juneau represents the southeast coastal area. Precipitation in Fairbanks under the RCP6 scenario exhibits considerable inter-decadal variation, with relative wet years during 1991–2015 and 2045–2055 and dry years during 2015–2045 and 2055–2100. After 2055, a persistent drought condition occurs, with annual precipitation around 0.3 m. Precipitation in Juneau is much larger than in the Interior, such as at Fairbanks, and displays an increasing trend during the period 1991–2100. Precipitation in Juneau during the 2090s, the final decade of the simulation, is around 3.5 m/year, compared to 3.0 m/year during the first decade, indicating an increase of about 17 %. Temperature exhibits significant increases under the RCP6 scenario for both interior and coastal locations in Alaska, particularly in winter (not shown).

Anomalies relative to the 15-year (1991–2005) averages of annual mean surface temperature, wind speed, and precipitation during three future decades (2030–2039, 2060–2069, and 2090–2099) indicate that, under the RCP6 scenario, the downscaled regional climate over Alaska and the surrounding oceans exhibits a significant warming trend (Fig. 7a–c). The strongest warming occurs over the Arctic Ocean, which is obviously attributable to the reduction in Arctic sea ice coverage (not shown). Temperature increases in Alaska are nonlinear, with relatively weak warming during the 2060s occurring between larger increases in the 2030s and 2090s.

The changes in surface wind speed show a strong inter-decadal variability (Fig. 7d–f). Over the Arctic Ocean stronger winds occur during the 2030s and 2090s, while weaker winds preside during the 2060s. A reduction in sea ice coverage favors the enhancement of surface winds (Stegall and Zhang 2012), but variations in the strength and location of synoptic weather patterns, such as the Beaufort High and low-pressure systems, can also impact surface winds (Zhang et al. 2013). Over the Bering Sea and most land areas in Alaska, decreased wind speeds are observed, though the magnitude of the decrease varies throughout the three future decades examined.

Enhanced precipitation emerges along the north and southeast coasts of Alaska and the west coast of Canada during each of the three decades, and over the Arctic Ocean during the 2030s and 2090s (Fig. 7g–i). Similar to the wind field changes seen over the Arctic Ocean, precipitation shows a slight decrease over the Arctic Ocean during the 2060s. On the other hand, over most land areas in Alaska, drier conditions prevail under the RCP6 scenario.

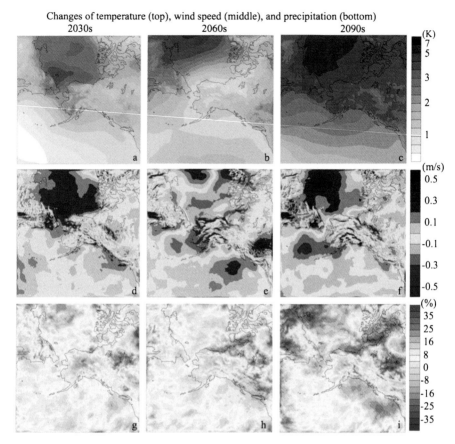

Fig. 7 Changes in annual mean surface temperature (**a**)–(**c**), wind speed (**d**)–(**f**), and precipitation (**g**)–(**i**) for three future decades (2030s, 2060s, and 2090s) relative to the 15-year (1991–2005) averages. Note that changes in precipitation are given as percentages

Summary

Dynamic downscaling of CMIP5/IPCC AR5 GCM simulations has been successfully conducted with a physically optimized version of WRF for Alaska and its surrounding areas. WRF downscaling of the twentieth century simulations from CCSM4 for the period 1991–2005 fundamentally captures reality. Strong seasonal variations are present in all three major surface climate parameters: temperature, wind speed, and precipitation. The downscaled twentieth century 15-year results are calibrated with in situ observations archived by the NCDC. A cold bias exists across most of the study area, except for a weak warm bias along the western and northern Alaskan coasts. In addition, downscaled winds are stronger than observations and precipitation is overestimated along the southeast coast. Following the algorithm developed by Zhang et al. (2007), the biases in the downscaled temperature, wind speed, and precipitation are correctable. Corrected downscaling results over Alaska and its surrounding areas indicate that, under the RCP6 climate change scenario, the study area will experience the following major changes:

- A significant warming trend, with strongest temperature increases over the Arctic Ocean due to a reduction in Arctic sea ice coverage
- Nonlinear temperature increases in Alaska, with relatively weak warming during the 2060s and strong warming during the 2030s and 2090s
- Strong inter-decadal variability in Arctic winds, with stronger winds during the 2030s and 2090s and weaker winds during the 2060s
- Decreased winds over the Bering Sea and most Alaskan land areas
- Enhanced precipitation along the north and southeast coasts of Alaska, the west coast of Canada, and over the Arctic Ocean
- Drier conditions over most land areas in Alaska

Acknowledgement This work was supported by the NSF Grants EAR-0943742, PLR-1304684, and ARC-1023592. Computing resources were provided by the Arctic Region Supercomputing Center at the University of Alaska Fairbanks.

References

Bengtsson, L., Botzett, M., & Esch, M. (1996). Will greenhouse gas-induced warming over the next 50 years lead to higher frequency and greater intensity of hurricanes? *Tellus, 48A*, 57–73.

Chen, F., & Dudhia, J. (2001). Coupling an advanced land-surface hydrology model with the PSU/NCAR MM5 modeling system. Part I: Model description and implementation. *Monthly Weather Review, 129*, 569–585.

Comiso, J. C., Parkinson, C. L., Gersten, R., & Stock, L. (2008). Accelerated decline in the Arctic sea ice cover. *Geophysical Research Letters, 35*, L01703.

Gardner, A. S., Moholdt, G., Cogley, J. G., Wouters, B., Arendt, A. A., Wahr, J., et al. (2013). A reconciled estimate of glacier contributions to sea level rise: 2003 to 2009. *Science, 340*, 852–857. doi:10.1126/science1234532.

Giorgi, F., Jones, C., & Asrar, G. (2009). Addressing climate information needs at the regional level: The CORDEX framework. *WMO Bulletin, 58*(V3), 175–183.

Grell, G. A., & Devenyi, D. (2002). A generalized approach to parameterizing convection combining ensemble and data assimilation techniques. *Geophysical Research. Letters, 29*, 1693. doi:10.1029/2002GL015311.

Gent, P. R., G. Danabasoglu, L. J. Donner, M. M. Holland, E. C. Hunke, S. R. Jayne et al. (2011). The community climate system model version 4. *Journal of Climate, 24*(19), 4973–4991.

Hock, R., de Woul, M., Radić, V., & Dyurgerov, M. (2009). Mountain glaciers and ice caps around Antarctica make a large sea-level rise contribution. *Geophysical Research Letters, 36*, L07501. doi:10.1029/2008GL037020.

Iacono, M., Delamere, J., Mlawer, E., Shephard, M., Clough, S., & Collins, W. (2008). Radiative forcing by long-lived greenhouse gases: Calculations with the AER radiative transfer models. *Journal of Geophysical Research, 113*, D13103.

Janjic, Z. I. (1994). The step-mountain Eta coordinate model: Further developments of the convection, viscous sublayer and turbulence closure schemes. *Monthly Weather Review, 122*, 927–945.

Janjic, Z. I. (1996). The Mellor-Yamada level 2.5 scheme in the NCEP Eta Model. In *11th Conference on Numerical Weather Prediction*, Norfolk, VA, American Meteorological Society, pp. 333–334.

Janjic, Z. I. (2002). Nonsingular implementation of the Mellor–Yamada level 2.5 scheme in the NCEP meso model. NCEP Office Note, No. 437, National Centers for Environmental Prediction, 61p.

Kalnay, E., et al. (1996). The NCEP/NCAR 40-year reanalysis project. *Bulletin of American Meteorological Society, 77*, 437–471.

Leung, L. R., Mearns, L. O., Giorgi, F., & Wilby, R. (2003). Workshop on regional climate research: Needs and opportunities. *Bulletin of American Meteorological Society, 84*, 89–95.

Lynch, A. H., McGinnis, D. L., & Bailey, D. A. (1998). Snow-albedo feedback and the spring transition in a regional climate system model: Influence of land surface model. *Journal of Geophysical Research, 103*, 29037–29049.

Mearns, L. O., Gutowski, W. J., Jones, R., Leung, L.-Y., McGinnis, S., Nunes, A. M. B., et al. (2009). A regional climate change assessment program for North America. *Eos, 90*, 311–312.

Mellor, G. L., & Yamada, T. (1982). Development of a turbulence closure model for geophysical fluid problems. *Review of Geophysics Space Physics, 20*, 851–875.

Morrison, H. C., Thompson, G., & Tatarskii, V. (2009). Impact of cloud microphysics on the development of trailing stratiform precipitation in a simulated squall line: Comparison of one- and two-moment schemes. *Monthly Weather Review, 137*, 991–1007.

Osterkamp, T. E. (2003). A thermal history of permafrost in Alaska. In *Proceedings of Eighth International Conference on Permafrost*, Zurich, pp. 863–868.

Skamarock, W. C., et al. (2008). A description of the advanced research WRF version 3. NCAR Technical Note, NCAR/TN–475+STR, 113pp.

Stegall, S. T., & Zhang, J. (2012). Wind field climatology, changes, and extremes in the Chukchi–Beaufort Seas and Alaska North Slope during 1979–2009. *Journal of Climate, 25*, 8075–8089.

Stroeve, J. C., Serreze, M. C., Holland, M. M., Kay, J. E., Malanik, J., & Barrett, A. P. (2012). The Arctic's rapidly shrinking sea ice cover: A research synthesis. *Climatic Change, 110*, 1005–1027.

Whitfield, J. (2003). Alaska's climate: Too hot to handle. *Nature, 425*, 338–339. doi:10.1038/425338a.

Zhang, J., Bhatt, U. S., Tangborn, W. V., & Lingle, C. S. (2007). Climate downscaling for estimating glacier mass balances in northwestern North America: Validation with a USGS benchmark glacier. *Geophysical Research Letters, 34*, L21505. doi:10.1029/2007GL031139.

Zhang, X., & Zhang, J. (2001). Heat and freshwater budgets and pathways in the Arctic Mediterranean in a coupled ocean/sea-ice model. *Journal of Oceanography, 57*, 207–237.

Zhang, X., Zhang, J., Krieger, J., Shulski, M., Liu, F., Stegall, S., et al. (2013). Beaufort and Chukchi Seas mesoscale meteorology modeling study, Final Report. U. S. Department of the Interior, Bureau of Ocean Energy Management. OCS Study BOEM 2013-0119, 204p., www.boem.gov/BOEM-2013-0119.

Part II
Fate and Transport of Contaminants

Application of Kalman Filter Embedded with Neural Network in 3-Dimensional Subsurface Contaminant Transport Modeling

Godwin Appiah Assumaning and Shoou-Yuh Chang

Abstract Predictive tools in the form of mathematical models have been used to simulate the movement and behavior of contaminants in groundwater. Conventionally, numerical models have been used to simulate these contaminants in the porous subsurface environment. A 3-D subsurface contaminant transport model numerically solved by approximation plagued the model with truncation and round-off errors and; assumes constant hydrologic parameters. In this research, to improve the accuracy of subsurface contaminant prediction spatially and temporally and to assess the impact of first-order decay rate parameter estimation, Kalman filter (KF) embedded with Neural Network (NN) was used in a specified 3-D domain space. The filter is perturbed with random Gaussian noise to reflect real life case of contaminant movement. Set of sparse observation points selected at specific locations are used to guide the filter at every time step to improve the accuracy of the prediction. The algorithms to generate the simulation results were run on Matlab 7.1. The accuracy of the KF embedded with NN, KF without Parameter Estimation and the numerical method were tested using Root Mean Square Error (RMSE) and Mean Absolute Error (MAE) equations. The KF embedded with NN performs better than both the discrete Kalman filter and the numerical method. Also, the KF embedded with NN is capable of reducing the error in the numerical solution by approximately 75 %.

Introduction

Accurate prediction of contaminant concentration and model parameters has become a major concern for hydrogeologist and environmental engineers. It helps in management decision-making concerning site remediation and water quality

G.A. Assumaning
Department of Civil and Environmental Engineering, NC A&T State University, Greensboro, NC 27411, USA

S.-Y. Chang (✉)
North Carolina A&T State University, Greensboro, NC, USA
e-mail: chang@ncat.edu

improvement. However, conventional methods such as numerical methods have been used to make predictions. These numerical model results are known to be less accurate due to the truncation, round-off, and approximation errors incorporated in them when they are derived and discretized (Chang and Latif 2010). The errors are due to assumptions made in the numerical model formulation. The model parameters are assumed to be constant in the simulation process. The objective of this work is to improve the accuracy and effectiveness of the three-dimensional (3-D) subsurface contaminant transport modeling using the KF embedded with NN with sparse observation data points and also to verify the importance of incorporating NN concept into the KF algorithm.

The KF is a set of mathematical equations that provides an efficient computational (recursive) means to estimate the state of a process, in a way that minimizes the mean of the squared error (Welch and Bishop 2004). It operates in two distinct phases: prediction and correction. Neural Networks (NN) are biologically inspired, i.e., they are composed of elements that perform in a manner that is analogous to the most elementary functions of the biological neuron (Rao 1999). The NN is coupled with KF to test its efficacy in contaminant prediction. Hendricks and Kinzelbach (2009) applied Ensemble Kalman filtering (EnKF) to off-line calibration of transient groundwater flow models with many nodes. Chang and Jin (2005) proposed the use of KF with regional noise to improve the accuracy of a contaminant transport models. A 3-D subsurface transport model was used by Cheng (2000) to generate the analytical, numerical, and KF results spatially and temporally under continuous contaminant input conditions. Leunga and Chan (2003) proposed the use of dual extended Kalman filtering (DEKF) to estimate the state of hidden layer, as well as the weights of the recurrent neural network. Kalman algorithm was used to estimate the state of the hidden layer. Li et al. (2002) presented a multistream Decoupled Extended Kalman filter (DEKF) training algorithm which provided efficient use of a parallel resource and more improved trained network weights. In this research, the first-order decay rate parameter is initially estimated using Extended Kalman filter (EKF) coupled with Neural Network. The contaminant concentration is then predicted at each time step with the estimated first-order decay rate parameter and compared to Kalman filter without Parameter Estimation and the numerical results. A Simulated True value is generated as reference data to test the accuracy of the results generated from all the techniques.

Methodology

Model description. The subsurface environment is made up of a complex, 3-D, heterogeneous, and hydrogeologic setting. Models are therefore formulated taking into account these processes. In this research, a 3-D subsurface advection-dispersion model for a nonconservative solute in a uniform, isotropic, saturated groundwater flow field along the x direction is used. The deterministic model for subsurface contaminant transport is represented in a partial differential equation (PDE) form in (1) (Cheng 2000).

$$\frac{\partial C}{\partial t} = \frac{D_x}{R}\frac{\partial^2 C}{\partial x^2} + \frac{D_y}{R}\frac{\partial^2 C}{\partial y^2} + \frac{D_z}{R}\frac{\partial^2 C}{\partial z^2} - \frac{V}{R}\frac{\partial C}{\partial x} - \frac{kC}{R} \quad (1)$$

where C is the concentration of contaminant, (mg/L), V is the linear velocity, (m/day), D_x, D_y, D_z is the dispersion coefficients in the x-, y-, and z-direction, respectively, (m^2/day), x, y, and z are the Cartesian coordinates, k is the first-order decay rate parameter, (1/day), t is time in days, and R is the retardation factor. The initial and boundary conditions of the subsurface transport model with instantaneous point source are given in (2) and (3), respectively.

$$C(x, y, z, t)_{t=0} = C_0(x, y, z) \quad (2)$$

$$C(x, y, z, t)_{t=\Omega} = 0 \quad (3)$$

Ω is chosen as the boundary.

Data assimilation with sparse observation data points. Practically, sparse observation data can be used in the data assimilation since it is quite expensive and laborious to take measurement at every location in the domain space. The domain space used in this research is represented as $(10 \times 10 \times 3)$ in a 3-D form with the nodes on x-direction, y-direction, and z-direction being 10, 10, and 3, respectively. Therefore, the total grid points in quasi 3-D form is 300, representing a full observation data set. Therefore, to simulate the model using the filters, 6 % of the full observation data points were used. This means 6 observation data points were selected from each layer for the filtering. Hence, a total of 18 observation points were chosen to run the simulation. The six observation nodes on the top layer (x–y plane) in the domain space were taken at locations (2,2), (5,2), (8,2), (2,8), (5,8), and (8,8).

Numerical method (FTCS). The 3-D subsurface contaminant transport model is numerically solved using a Finite-differencing scheme called Forward-Time and Central-Space (FTCS). The space and time steps chosen satisfy the numerical stability and convergence criteria of the Peclet number. Jin (1996) and Chang and Assumaning (2011) proposed the use of FTCS in their contaminant modeling approach using the Kalman filter and Particle filter. The convergence and stability criteria for the scheme were adhered to in the discretization and implementation process. The solution of the numerical method in a matrix form is given as:

$$X_{t+1} = AX_t \quad (4)$$

where X_{t+1} is the vector of contaminant concentration at all nodes at time, $t+1$, X_t is the vector of contaminant concentration at all nodes at time, t, A is the State Transition Matrix (STM) containing the parameters for the model. The algorithm for the numerical scheme was coded in Matlab 7.1 to estimate the concentration of the contaminant.

Process and observation models. Two main data sets are required to run the data assimilation filters. The two governing equations needed to generate these data sets are system and observation equations. These two equations are dynamic and stochastic in nature. The process or system model in this research is the numerical model with an additional Gaussian noise. The process equation is a discrete-time controlled process (Cheng 2000). The generic form of the process equation is given as

$$X_{t+1} = AX_t + w_t \quad t = 0, 1, 2, 3, \ldots \quad (5)$$

where w_t is the system noise vector assumed to be normally distributed with covariance of Q_t and a zero mean. A standard deviation of 10 % was added to the numerical scheme to generate the dynamic system states. The observation/measurement data set is generated from the Simulated True value with additional random Gaussian noise. The equation governing the observation data is given as

$$Z_t = HX_t^T + O_t \quad (6)$$

where Z_t is the vector of the observed values for all nodes at time step t, X_t^T is the Simulated True value of the state for all nodes at time step t, O_t is vector of the observation error, and H is the measurement sensitivity matrix. The observation error vector O_t is assumed to be normally distributed with covariance of R_t and zero mean. A standard deviation of 5 % was chosen and added to the Simulated True value to generate the observation data set.

Kalman filter without parameter estimation. The KF is considered to be a very powerful tool since it supports estimation of past, present, and even future states even when the precise nature of the modeled system is unknown (Welch and Bishop 2004). The solution is recursive since updated estimate of the state (concentration) is computed from the previous estimate and new input data (Haykin 2001). In this study, KF is used to estimate the state (concentration) of the contaminant in a porous subsurface environment. This data assimilation is carried out primarily to reduce the variance estimates of the states. The KF estimation equation depicting the stochastic condition is given as:

$$X_t(+) = X_t(-) + K_t[Z_t - HX_t(-)] \quad (7)$$

where $X_t(+)$ is the vector of estimated states after the KF adjustment, $X_t(-)$ is the vector of estimated states before the KF adjustment, and K_t is the Kalman gain matrix. The K_t is derived by minimizing the trace of the posterior error covariance matrix, $P_t(+)$. The trace is minimized when the derivative of $P_t(+)$ is zero. K_t determines how much the estimated value using KF can gain from the observation. The K_t is determined by:

$$K_t = P_t(-)H^T \left(HP_t(-)H^T + R_t\right)^{-1} \quad (8)$$

$(\cdot)^T$ and $(\cdot)^{-1}$ denotes the transpose and inverse of matrix, respectively, in (8). The posterior and prior error covariance matrices given in (9) and (10), respectively, are advanced recursively for all time steps by:

$$P_t(+) = (I - K_t H) P_t(-) \quad (9)$$

$$P_{t+1}(-) = A P_t(+) A^T + Q_t \quad (10)$$

The initial value of the posterior error covariance matrix is given as

$$P_0 = E\left[(X_0 - E[X_0])(X_0 - E[X_0])^T\right] \quad (11)$$

where Q_t is the system noise covariance matrix and X_0 is the estimated state at time step zero.

Kalman filter embedded with neural network. Walker (2006), Sen et al. (2004), and Puskorius and Feldkamp (1997) have applied KF embedded with NN in their respective works. In this research, the measurement sensitivity (derivative) matrix H is obtained by feed forward layered NN algorithm, the first-order decay rate parameter k and trainable weights are estimated using Extended KF (EKF) coupled with NN algorithm. KF is then applied to estimate the concentration at each time step.

EKF has been used in weight training since weight updates are based on second-order derivative information as compared to back propagation which uses first-order derivative information. EKF also has a stochastic component in its weight update process. The algorithm seeks to find the minimum weight values that minimize the sum of squared error. The nonlinear dynamical weight-space models and decay rate parameter-space models are provided as

$$b_{t+1} = b_t + w_t \quad (12)$$

$$k_{t+1} = k_t + w_t \quad (13)$$

$$y_t = h(b_t, u_t, k_t) + O_t \quad (14)$$

where (12), (13), and (14) are process equations for the trainable weights and decay rate parameter; and measurement equation, respectively. Also, u_t is the input parameter, w_t and O_t are independent zero-mean white Gaussian noise processes with covariance matrices Q_k and R_k, respectively. The measurement sensitivity (derivative) matrix H is given as

$$H_t = u_t^i \left(\widetilde{b}_t^i\right)^T \quad (15)$$

where u_t^i is the ith node's input vector and $\widetilde{b}_t^i$ is a vector of partial derivatives of the network's outputs with respect to the ith node's net input. The following recursive and dynamic equations are also used

$$A_t = \left[R_t + H_t^T P_t H_t\right]^{-1} \tag{16}$$

$$K_t = P_t H_t A_t \tag{17}$$

$$b_{t+1} = b_t + K_t(y_t - \widetilde{y}_t) \tag{18}$$

$$k_{t+1} = k_t + K_t(y_t - \widetilde{y}_t) \tag{19}$$

$$P_{t+1} = P_t - K_t H_t^T P_t + Q_t \tag{20}$$

where A_t is the global scaling matrix, b_{t+1} is the vector of estimated network's weight parameter values at time, $t+1$, b_t is the vector of network's weight parameter values at time, t, k_{t+1} is the vector of estimated first-order decay rate parameter at time, $t+1$, k_t is the vector of first-order decay rate parameter at time, t, $\widetilde{y}_t$ is the network's output vector at time, t, y_t is the target weight vector at time, t. The initial weight values are randomly picked from zero-mean normal distribution and the first-order decay rate parameter is initialized with the value used in the numerical scheme. Typically, the initial error covariance is provided as

$$P_o = \varepsilon^{-1} I \tag{21}$$

where ε is 0.01 for sigmoidal activation function, and I is an identity matrix. The estimated parameters are updated at each time step. After the network weights and the first-order decay rate parameter are estimated, (5)–(11) is applied to estimate the contaminant concentrations.

Testing the prediction techniques results. The measure of variability in the predicted values is obtained by using two methods. These methods are Root Mean Square Error (RMSE) and Mean Absolute Error (MAE). The RMSE and MAE equations are defined in (22) and (23), respectively.

$$\text{RMSE}(t) = \sqrt{\frac{1}{N}\sum \left[C^E(x,y,z,t) - C(x,y,z,t)\right]^2} \tag{22}$$

$$\text{MAE}(t) = \sum \frac{|C^E(x,y,z,t) - C(x,y,z,t)|}{N} \tag{23}$$

where N is the number of sampling nodes, C^E is the estimated concentration of the contaminant at time, t and C is the Simulated True value at time, t.

Results and Discussion

Model parameters. The results of the filters and the numerical scheme were obtained by setting the hydrogeologic parameters needed to perform the simulations. An instantaneous contaminant was injected into the grid point at coordinates (5,5,1) with concentration of 10,000 mg/L. The total number of grid points is

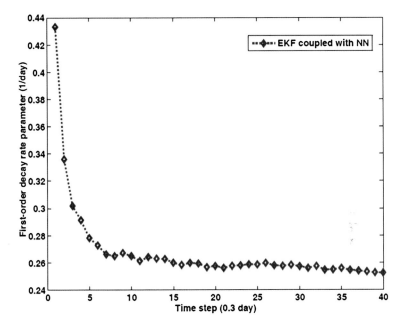

Fig. 1 Profile of first-order decay rate parameter estimation using EKF coupled with NN

300, sparse observation points is 18, first-order decay rate parameter is 0.35 1/day, linear velocity is 0.15 m/day, retardation factor is 1.125, dispersion in x-, y-, and z-direction is 0.35 m^2/day, 0.35 m^2/day, and 0.35 m^2/day, respectively. The grid interval is 2 m in all directions. Simulation time is 12 days, time interval is 0.3 day, and a total time step is 40.

Decay rate parameter estimation using extended Kalman filter coupled neural network. The estimated values of the decay rate parameter are shown in Fig. 1 for each time step. The profile stabilizes and converges at time step 15 with an estimated value of approximately 0.25 1/day. The convergence value is an indication that the actual first-order decay rate parameter is 0.25 1/day. The profile also shows the reduction in estimation error with time. The estimation is facilitated by the introduction of set of measurement. The maximum value 0.43 1/day estimated at the beginning of the process is as a result of the learning process of the NN and the data assimilation process. The estimated parameter values are subsequently introduced into the system model of the KF to facilitate the prediction of the contaminant concentration.

Comparison of all prediction techniques results. To analyze the accuracy of each prediction results, the contour of all techniques results for layer 1 at time step 40 were plotted as shown in Fig. 2. From Fig. 2, numerical solution is farther away from the Simulated True value. This indicates that the error in the numerical solution is relatively the largest compared to other prediction techniques. The closest prediction technique to the Simulated True value is KF embedded with NN. The deviation by the technique from the Simulated True value is very minimal. This suggests that the KF embedded with NN is relatively better than the rest of the techniques used in this research.

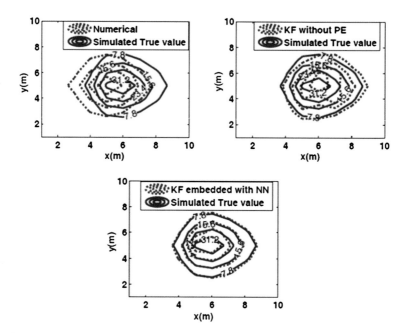

Fig. 2 Contaminant concentration contours for all the prediction techniques at time step 40 for layer 1

Accuracy of the numerical method and data assimilation filters. The RMSE profiles for all prediction techniques can be seen in Fig. 3. The numerical solution shows the maximum error at all the time steps. The filters were found to be unstable in nature due to fewer observation data used and the random Gaussian noise introduced into the data assimilation process. The randomness in profile also explains the heterogeneity of the real field. Among the filters, it can be established that KF embedded with NN profile is relatively stable and converges faster than the other filters. RMSE of KF without parameter estimation profile is highly erratic in nature and therefore converges slowly with time. Although, the numerical solution converges, the filters were found to converge faster after time step 15. The maximum error for KF without parameter estimation and KF embedded with NN are about 38 mg/L and 26 mg/L, respectively. The maximum error in the numerical scheme was found at time step 7 with an error of about 37.5 mg/L. The KF embedded with NN performs better than the other filters and the numerical method due to its relatively smallest error at each time step. At the end of the simulation, errors in the numerical solution and the filters were approximately about 4 mg/L and 1 mg/L, respectively. This indicates that the filters are capable of reducing the error in the numerical method by 75 % at the end of the prediction.

Another statistical means of measuring variability in the prediction results was adopted. The MAE used in this work to find the absolute difference between the predicted and the Simulated True value at every time step. Figure 4 shows the profile of the MAE profiles for all prediction techniques. The deviation by

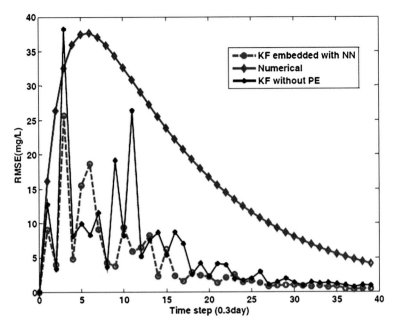

Fig. 3 Root mean square error (RMSE) profiles for all prediction techniques

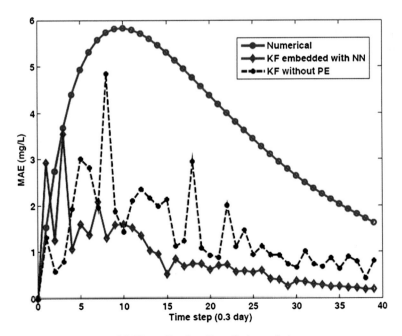

Fig. 4 Mean absolute error (MAE) profiles for all prediction techniques

the numerical solution from the data assimilation filters is very visible and indicates the degree of error in the results. The profile also shows the erratic nature of the filters from time 1–20. The randomized nature of the filters is as a result of the stochastic Markov chain process used in simulating the contaminant transport in a heterogeneous subsurface environment. The KF embedded with NN performs better than other prediction techniques. At the end of the simulation, errors in the numerical solution and the filters were approximately about 1.8 and 0.4 mg/L, respectively. This indicates that the filters are capable of reducing the error in the numerical method by 77 % at the end of the prediction.

Conclusion

The prevention of groundwater quality deterioration has led to the development of models to analyze and predict the movement of contaminants in the subsurface environment. In this research, data assimilation filters namely KF without Parameter Estimation and KF embedded with NN were used as tools for predicting contaminant concentration in subsurface porous environment. The Parameter Estimation technique was also adopted due to the uncertainties associated with hydrogeologic parameters and aquifer properties.

The data assimilation filters show a strong convergence trends in the error estimation profiles given. They are initially erratic but subsequently stabilize and converge with time indicating the effectiveness of the filtering process. The filters have an advantage of minimizing the error or residual between the observed values and the estimated values. From the RMSE profile, the errors in the numerical solution and the filters were approximately about 4 and 1 mg/L, respectively, indicating a 75 % reduction in the numerical error if the filters are adopted.

The KF embedded with NN was found to perform better than the other prediction techniques used. The incorporation of EKF coupled with NN for Parameter Estimation positively impacted on the prediction accuracy of the contaminant concentration. All the profiles on RMSE and MAE confirm the higher accuracy of the KF embedded with NN. The adjusted R square value was 98 % which is higher than the KF without Parameter Estimation of 93.5 %.

The filters also proved to be efficient when run with sparse observation data points as opposed to full observation data used in previous research works on KF. The contaminant transport simulations for all the data assimilation filters and numerical method were run on a personal computer (PC) with processor speed of 2.99 GHz and RAM of 3.25 GB. The KF with Parameter Estimation take about 8 min to run while the numerical method takes 3 min. The filters have high computational time and challenges due to the Parameter Estimation, prediction, and updating processes of the algorithm. However, it is worth using the filters due to their effectiveness and accuracy in prediction.

Acknowledgments This work was sponsored by the Department of Energy Samuel Massie Chair of Excellence Program under Grant No. DF-FG01-94EW11425. The views and conclusions contained herein are those of the writers and should not be interpreted as necessarily representing the official policies or endorsements, either expressed or implied, of the funding agency.

References

Chang, S. Y., & Assumaning, G. (2011). Subsurface radioactive contaminant modeling using particle and Kalman filter schemes. *Journal of Environmental Engineering, 137*(4), 221.

Chang, S. Y., & Jin, A. (2005). Kalman filtering with regional noise to improve accuracy of contaminant transport models. *Journal of Environmental Engineering, 131*(6), 971–982.

Chang, S. Y., & Latif, S. M. I. (2010). Extended Kalman filtering to improve the accuracy of a subsurface contaminant transport model. *Journal of Environmental Engineering, 136*(5), 466–474.

Cheng, X. (2000). Kalman filter scheme for three-dimensional subsurface transport simulation with a continuous input. Master of Science Thesis, Civil and Environmental Engineering. Greensboro, NC: North Carolina Agricultural and Technical State University.

Haykin, S. (2001). *Kalman filtering and neural networks*. New York: Wiley. ISBN 0-471-36998-5.

Hendricks, F. H. J., & Kinzelbach, W. (2009). Ensemble Kalman filtering versus sequential self-calibration for inverse modeling of dynamic groundwater flow systems. *Journal of Hydrology, 365*, 261–274.

Jin, A. (1996). An optimal estimation scheme for subsurface contaminant transport model using Kalman-Bucy filter. Master's Thesis, Department of Civil and Environmental Engineering. Greensboro, NC: North Carolina A&T State University.

Leunga, C. S., & Chan, L. W. (2003). Dual extended Kalman filtering in recurrent neural networks. *Neural Networks, 16*, 223–239.

Li, S., Wunsch, D. C., O'Hair, E., & Giesselmann, M. G. (2002). Extended Kalman filter training of neural networks on a SIMD parallel machine. *Journal of Parallel and Distributed Computing, 62*, 544–562.

Puskorius, G. V., & Feldkamp, L. A. (1997). Multi-stream extended Kalman filter training for static and dynamic neural networks. *IEEE International Conference on Systems, Man and Cybernetics, 3*, 2006–2011.

Rao, A. S. (1999). Artificial neural network embedded Kalman filter bearing only passive target tracking. In *Proceedings of the 7th Mediterranean Conference on Control and Automation (MED99)* Haifa, Israel, June 28–30, 1999.

Sen, Z., Altunkaynak, A., & Ozger, M. (2004). Sediment concentration and its prediction by perceptron Kalman filtering procedure. *Journal of Hydraulic Engineering, 130*(8), 816–826.

Walker, D. M. (2006). Parameter estimation using Kalman filters with constraints. *International Journal of Bifurcation and Chaos, 16*(4), 1067–1078.

Welch, G., & Bishop, G. (2004). *An introduction to the Kalman filter*. Chapel Hill: Department of Computer Science, University of North Carolina at Chapel Hill.

Application of Adaptive Extended Kalman Filtering Scheme to Improve the Efficiency of a Groundwater Contaminant Transport Model

Shoou-Yuh Chang and Elvis B. Addai

Abstract Pollution of groundwater can be harmful to the environment. The use of subsurface contaminant transport models, combined with stochastic data assimilation schemes, can give on-target predictions of contaminant transport to enhance the reliability of risk assessment in the area of environmental remediation. In this study, a two-dimensional transport model with advection and dispersion is used as the deterministic model of contaminant transport in the subsurface. An Adaptive Extended Kalman Filter (AEKF) is constructed as a stochastic data assimilation scheme to meliorate the prediction of the contaminant concentration. The effectiveness of the AEKF is determined by using a root mean square error (RMSE) of pollutant concentrations in contaminant transport modeling. The implementation of the AEKF was successful in improving the prediction accuracy of the deterministic model by about 60.7 % which shows a substantial improvement in the prediction of the contaminant concentration in the subsurface environment.

Introduction

Information on the movement and behavior of contaminants in the subsurface at polluted location is necessary to comprehend the nature of the existing problem and the current or potential public health or ecological risks in order to put into place site-specific cleanup goals that are viable and to design a remediation program that is affordable, reliable and likely to achieve the cleanup goals. Knowledge of groundwater contaminant is required to avoid pumping contaminated water for human consumption. Mathematical models have been recognized as the most

S.-Y. Chang (✉)
North Carolina A&T State University, Greensboro, NC, USA
e-mail: chang@ncat.edu

E.B. Addai
Department of Civil Engineering, North Carolina Agricultural & Technical State University, Greensboro, NC 27411, USA
e-mail: ebaddai@ncat.edu

effective tool for explaining how a plume of contaminant evolves. Subsurface contaminant transport models can describe the spatial and temporal distribution of contaminant as well as their movement to potential receptor points (Tam and Beyer 2002). A mathematical formulation and some numerical approximation techniques are described for a system of coupled partial differential and algebraic equations describing multiphase flow, transport, and interactions of chemical species in the subsurface (Todd et al. 1996). Although many transport problems must be solved numerically, analytical solutions are still pursued by many scientists because they can provide better physical insights into the problems. Analytical solutions are usually derived from the basic physical principles and free from numerical dispersions and other truncation errors that often occurred in numerical simulations (Zheng and Bennett 1995). Lin et al. (2010) developed a simplified numerical model of groundwater and solute transport. In another study, Park and Zhan (2001) investigated an analytical solution of contaminant transport from finite one-, two-, and three-dimensional sources in a finite-thickness aquifer. A three-dimensional numerical model for groundwater flow and heat transport is used to analyze the heat exchange in the ground (Lee 2011). Gerald (2011) applied the finite difference method to a partial differential equation. The partial differential equation is replaced with a discrete approximation in the numerical solution. The backward time, centered space (BTCS) was formulated and applied to one-dimensional heat equation. Chang and Jin (2005) improved the efficiency of a contaminant transport models by using Kalman filter with regional noise. The Kalman filter can be used for past, present, and even future state estimates (Welch and Bishop 2006). Chang and Latif (2010) applied Extended Kalman filter to improve the accuracy of a subsurface contaminant transport model. Han et al. (2009) presented an Adaptive Extended Kalman filter method to estimate the state-of-charge. The Adaptive Extended Kalman filter reduced the state-of-charge estimation error when working with the unknown process and measurement noise covariance values.

An Adaptive Extended Kalman filter (AEKF) scheme is constructed by regulating the system error covariance and the optimal Kalman gain in order to reduce the effect of errors in the process and observation noise statistics during filter operation. The adaptive filtering technique which is based on the Innovation Covariance Scaling and Gain Correction (ICS-GC) addresses the effect of unaccounted errors in the process and observation models (Kim and Lee 2006). The numerical solution was discretized using backward in time and centered in space (BTCS). The effectiveness and the performance of the proposed AEKF algorithm have been experimentally tested using root mean square error. The objectives for this research are to improve the prediction of contaminant plume using AEKF in a subsurface environment and to examine the performance of the AEKF in the two-dimensional contaminant transport model.

Methodology

A two-dimensional subsurface advection-dispersion model for the transport of contaminant in the horizontal plane ($x-y$) and advection in the x direction was used to examine the accuracy and effectiveness of the numerical, the Extended Kalman filter (EKF) and the AEKF results relative to the analytical solution that has been randomized (reference true value). The advection-dispersion equation for a two-dimensional transport in the $x-y$ plane is characterized by the following partial differential equation:

$$\frac{\partial C}{\partial t} = \frac{D_x}{R}\left(\frac{\partial^2 C}{\partial x^2}\right) + \frac{D_y}{R}\left(\frac{\partial^2 C}{\partial y^2}\right) - \frac{V}{R}\left(\frac{\partial C}{\partial x}\right) - \frac{k}{R}C \tag{1}$$

where C = concentration of contaminant in the solute phase (mg/L); V = linear velocity in the x-direction (m/d); t = time (day); R = retardation factor; k = first-order decay rate (1/day); D_x, D_y = dispersion coefficients in the x and y directions, respectively, (m^2/d); x, y = Cartesian coordinates (m). In order to express the deterministic model in a more convenient form for the implementation of the data assimilation schemes, a Backward-Time and Central-Space (BTCS) differencing scheme was developed to solve the two-dimensional transport model numerically. The solved two-dimensional transport model can be rewritten in a state-space form as

$$\mathbf{x}_t = \mathbf{A}\mathbf{x}_{t-1} \tag{2}$$

where $\mathbf{x}_t$ = the vector of contaminant concentration at all nodes at time, t; $\mathbf{x}_{t-1}$ = the vector of contaminant concentration at all nodes at time, $t-1$; $\mathbf{A}$ = the State Transition Matrix (STM) containing the parameters for the model which advances the current state to the next time step.

Transport processes can be simulated by stochastic data assimilation schemes with uncertain sources and inaccurate transport parameters by introducing a random noise term in the deterministic dynamics (Saad 2007). The process equation is given as

$$\mathbf{x}_t = \mathbf{A}\mathbf{x}_{t-1} + \mathbf{p}_t, \; t = 0, 1, 2, 3 \ldots \tag{3}$$

where $\mathbf{x}_t$ = vector of contaminant concentration at all nodes at time, t; $\mathbf{x}_{t-1}$ = vector of contaminant concentration at all nodes at time, $t-1$; $\mathbf{A}$ = State Transition Matrix (STM) that runs till the last time step; $\mathbf{p}_t$ = model system error or process noise. The system model error, $\mathbf{p}_t$ is assumed to have zero mean and covariance matrix, $\mathbf{Q}_t$. The model system error is the difference between the optimal estimate of the true state and the model prediction. The equation governing the observation data for the entire domain in this study is given as

$$Z_t = HX_t^T + O_t \tag{4}$$

where Z_t = state vector for observed values for all nodes at time step t; H = measurement sensitivity matrix; X_t^T = the transpose of the true optimal estimate of the state; O_t = vector of the observation error. The observation noise, O_t is assumed to have zero mean and covariance matrix R_t. H is constructed as an identity matrix with n being the number of nodes in the model domain, in this study the number of nodes were chosen to be 625.

In order to implement the data assimilation schemes, observation information is required to guide the system model to project the optimal contaminant concentration. The analytical solution which has been randomized governing the advection-dispersion partial differential equation is given by

$$C(x,y,t) = \frac{M_0}{4\pi b\eta t \sqrt{D_x D_y}} e^{\left(-\frac{(x-Vt/R)^2}{4D_x t/R} - \frac{y^2}{4D_y t/R} - kt\right)} + w_t \tag{5}$$

where C = concentration of contaminant in the solute phase (mg/L); D_x = dispersion coefficients in x directions (m²/d); D_y = dispersion coefficients in y directions (m²/d); η = porosity of the medium; b = aquifer thickness (m); R = retardation factor (dimensionless); V = linear velocity in the x-direction (m/d); k = the first-order contaminant decay rate (1/day); M_0 = instantaneous mass input (g); x = Cartesian coordinates (m) in the x direction; y = Cartesian coordinates (m) in the y direction; t = time (day); w_t = {0, 5 %}. Equation (5) represents the variations in time and distance in concentration, depending on the initial mass of contaminant per unit area injected across the aquifer cross section during a spill at time $t = 0$. An observation was created by introducing an observation error of 2.5 % into the true solution.

The EKF is the nonlinear version of the Kalman filter which linearizes about an estimate of the current mean and covariance. It provides a consistent first-order approximation to the optimal estimate of the state and of the time-dependent model uncertainties, both when data are available and when they are not (Kao et al. 2003). A Kalman filter that linearizes about the current mean and covariance is referred to as an EKF (Welch and Bishop 2006). The two nonlinear dynamical state–space models are

$$x_{t+1} = A(t, x_t) + p_t \tag{6}$$

$$z_t = h(t, x_t) + o_t \tag{7}$$

The functional $A(\mathbf{t}, \mathbf{x}_t)$ and $h(\mathbf{t}, \mathbf{x}_t)$ represent nonlinear transition matrix function and nonlinear measurement matrix that are time-variant, respectively. The nonlinear matrices are transformed into partial derivatives given as

$$F_{t+1} = \frac{\partial A(t, x_t)}{\partial x} \qquad (8)$$

$$H_t = \frac{\partial h(t, x_t)}{\partial x} \qquad (9)$$

where F_{t+1} and H_t are the Jacobian matrices for the state transition and the measurement, respectively. Using the basic idea of Kalman Filtering, the transformed state transition and measurement matrix is then used which makes the filter to work in the extended form.

An AEKF is envisioned where changes are made during each iteration to the state model covariance to improve convergence. It is expected that the execution of an AEKF will help to reduce the effect of unaccounted errors in the process and observation noise covariance. The adaptive filtering technique which is based on the Innovation Covariance Scaling and Gain Correction (ICS-GC) addresses the effect of unaccounted errors in the process and observation models (Kim and Lee 2006). The algorithm for the AEKF is given as

Prediction stage:

$$\mathbf{x}_t(-) = \mathbf{A}\mathbf{x}_{t-1} + \mathbf{p_t} \qquad (10)$$

$$\overline{\mathbf{P}}_t(-) = \alpha_t \left[\mathbf{A}\mathbf{P}_{t-1}\mathbf{A}^T + \mathbf{Q}_t \right] \qquad (11)$$

Update stage:

$$\overline{\mathbf{K}}_t = \overline{\mathbf{P}}_t(-)\mathbf{H}^T \left(\mathbf{H}\overline{\mathbf{P}}_t(-)\mathbf{H}^T + \lambda_t \mathbf{R}_t \right)^{-1} \qquad (12)$$

$$\mathbf{x}_t(+) = \mathbf{x}_t(-) + \overline{\mathbf{K}}_t [\mathbf{z}_t - \mathbf{H}\mathbf{x}_t(-)] \qquad (13)$$

$$\overline{\mathbf{P}}_t(+) = (\mathbf{I} - \overline{\mathbf{K}}_t \mathbf{H}) \overline{\mathbf{P}}_t(-) \qquad (14)$$

where $\mathbf{x}_t(-)$ = the estimated state before the AEKF adjustment with covariance, $\mathbf{P}_t(-)$; $\mathbf{x}_t(+)$ = the estimated state after the AEKF adjustment with covariance, $\mathbf{P}_t(+)$; $\mathbf{I}$ = an identity matrix.

$\overline{\mathbf{K}}_t$ = the optimal Kalman gain matrix which minimizes the optimal estimate error covariance matrix, $\mathbf{P}_t$; λ_t = Forgetting factor at time step t; α_t = Adaptive factor at time step t.

The model predictions are compared with a simulated true solution to estimate the error parameter (RMSE). The RMSE indicates the degree of error in each result in relation to the true solution. The RMSE is given as

$$\text{RMSE}(t) = \sqrt{\frac{1}{N-1} \sum \left[C^T(x, y, t) - C^P(x, y, t) \right]^2 }$$

where RMSE (t) = Root Mean Square Error (mg/L) at time step, t; N = Number of sampling nodes; $C^P(x, y, t)$ = Predicted value of concentration at node (x, y) at time step, t; C^T = Reference true value of concentration at node (x, y) at time step, t.

Results and Discussion

The simulation was conducted using a 37.5 m by 37.5 m domain space with grid points of 625 on a two-dimensional plane. The grid intervals used in this study in the two-dimensional plane in the x direction, Δx is 1.5 m and that of the y direction, Δy is also given as 1.5 m. The parameter values used in the subsurface to simulate the contaminant transport were acquired from Zou and Parr (1995). The dispersion coefficients, D_x, in x directions and the dispersion coefficients, D_y, in y directions are 1.554 m²/d and 0.4662 m²/d, respectively. The time adopted for each time step, the thickness of the aquifer, and the porosity, η, in the analytical equation were taken as 0.2 day, 6.1 m and 0.3, respectively. The retardation factor, R, is assumed to be 1.2 while the linear velocity, V, is taken to be 1.5 m/d. The initial concentration of contaminant injected into the grid point at coordinates (5, 10) was 10,000 mg/L and the instantaneous contaminant mass at the grid point is 1604 g. Simulations were carried out by running Matlab codes for numerical, true solution, EKF and AEKF. Figures 1, 2, and 3 show the contour profile of the contaminant plume prediction of the deterministic model, EKF and AEKF in comparison with

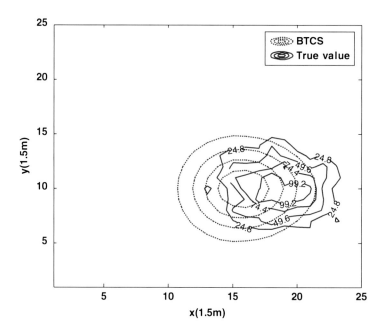

Fig. 1 Comparison of numerical results and true value at time step 50

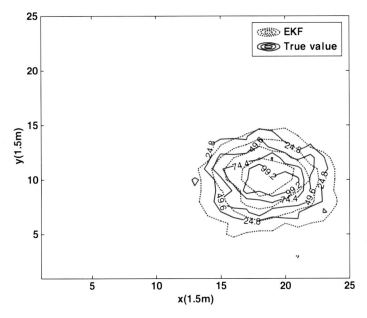

Fig. 2 Comparison of EKF results and true value at time step 50

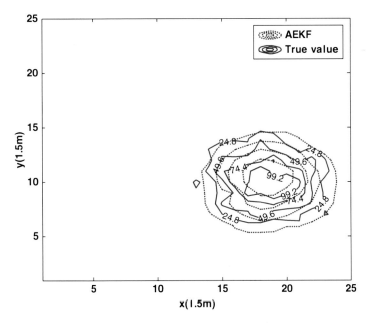

Fig. 3 Comparison of AEKF results and true value at time step 50

the simulated true solution, respectively. Many factors combine to determine the quality and usefulness of a model predication. Numerical methods involve approximations which introduce errors into the deterministic model thereby affecting the accuracy and precision of the prediction. The setting of stability and convergence criteria in the results generation is another form of error in the numerical approach. Moreover, due to the heterogeneous nature of the subsurface, hydraulic parameters and initialization data acquired from the field may vary, as a result assumptions are made on the parameters and the model used in estimation which further introduce error into the numerical method. The EKF remains a popular choice that has been applied in various fields. However, the EKF is vulnerable to linearization errors, which can lead to poor performance or divergence (Guoquan and Stergios 2008). Also, the EKF assumes a complete knowledge of the process and observation noise covariance matrices which may be estimated inaccurately and causes the filter to diverge (Hide et al. 2004). Therefore, the uncertainties in the process and observation noise covariance are addressed by introducing the adaptive filtering technique into the EKF scheme. The AEKF as a data assimilation scheme combines observation data and model dynamics to give an improved prediction of the state at discrete time and spatial points. The AEKF reduces the effect of inaccuracies in the initial process noise covariance by assigning less weight to the system model (Hu et al. 2003). The closeness of the AEKF to the true solution is an indication of improvement in the system model by the introduction of the adaptive technique thereby providing a more accurate prediction. The AEKF produces faster and smoother convergence than the EKF. The AEKF works better than the deterministic model and the EKF in terms of prediction accuracy.

Figure 4 shows the mesh profile at time step 50, indicating the peak concentration of the true solution, the AEKF, EKF, and the BTCS.

At the end of the simulation, the true solution peak concentration was approximately 99.2 mg/L, which was closely predicted by the AEKF and the EKF to be 110 mg/L and 150 mg/L, respectively, whereas the BTCS predicted relatively higher peak concentration of 240 mg/L. The AEKF is closer to the true solution than both the EKF and the BTCS indicating that the initialization of the adaptive filtering scheme was successful in bringing down the prediction error of the EKF scheme.

The effectiveness of AEKF, EKF, and numerical results were measured using root mean square error (RMSE). Figure 5 shows the RMSE profile for the approaches that were used in this study. An average prediction error of approximately 29.9, 21.3, and 18.2 mg/L were obtained for the deterministic model, the EKF and the AEKF, respectively, for the entire simulation period when compared with the true solution. At the end of the simulation, the prediction error for the BTCS was 28 mg/L while the EKF and the AEKF were reduced to 18 and 11 mg/L in the order given. The EKF improved the prediction by an average of 28.8 % relative to BTCS while the AEKF improved the prediction by an average of 39.1 % relative to BTCS for the entire simulation period. The adaptive factor and forgetting factor used in the Innovation Covariance Scaling and Gain Correction (ICS-GC)

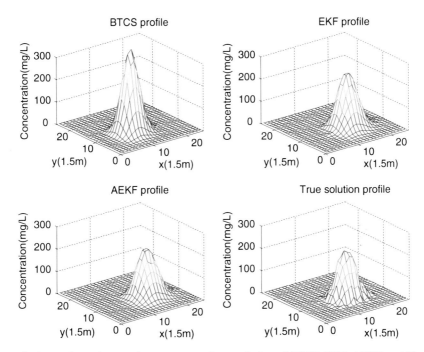

Fig. 4 Comparison of contaminant concentration prediction of BTCS, EKF, AEKF, and True solution at time step 50

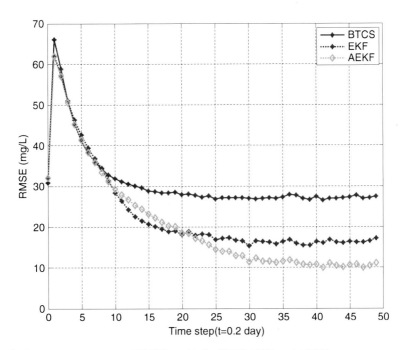

Fig. 5 Root mean square error (RMSE) profile for BTCS, EKF, and AEKF

adaptive scheme by tuning the system error covariance and the optimal Kalman gain improves the accuracy of the filter prediction. From Fig. 5, it is evident that the filters gave more accurate estimation than the deterministic model.

Conclusion

Despite the fact that numerical model is useful in predicting the fate and transport of contaminants in groundwater, the complex nature of the subsurface environment makes it extremely difficult to give accurate prediction. The deterministic model assumes constant parameters and homogeneity in generating results, while both the EKF and the AEKF takes into consideration updating and prediction in results generation. The introduction of the EKF improved the BTCS prediction by reducing the model error from 28 to 18 mg/L, thus improving the prediction accuracy by about 35.7 %. The implementation of the AEKF was successful in improving the prediction accuracy of the BTCS model by about 60.7 %.

Acknowledgement This work was sponsored by the Department of Energy Samuel Massie Chair of Excellence program under grant number DE-NA0000718.

References

Chang, S. Y., & Jin, A. (2005). Kalman filtering with regional noise to improve accuracy of contaminant transport models. *Journal of Environmental Engineering, 131*(6), 971–982.
Chang, S. Y., & Latif, S. M. I. (2010). Extended Kalman filtering to improve the accuracy of a subsurface contaminant transport model. *Journal of Environmental Engineering, 136*(5), 466–474.
Gerald, W. R. (2011). Finite difference approximations to the heat equation. Port-land, OR: Mechanical Engineering Department Portland State University. Retrieved August 15, 2012, from www.f.kth.se/~jjalap/numme/FDheat.pdf.
Guoquan, P. H., & Stergios, I. R. (2008). An observability constrained UKF for improving slam consistency. *Multiple Autonomous Robotic Systems Laboratory*, Technical Report Vol. 2, pp. 1–29.
Han, J., Kim, D., & Sunwoo, M. (2009). State-of-charge estimation of lead-acid batteries using an adaptive extended Kalman filter. *Journal of Power Sources, 188*, 606–612.
Hide, C., Michaud, F., & Smith, M. (2004). Adaptive Kalman filtering algorithms for integrating GPS and low cost INS. *IEEE Position Location and Navigation Symposium*, Monterey, CA, 227–233.
Hu, C., Chen, W., Chen, Y., & Liu, D. (2003). Adaptive Kalman filtering for vehicle navigation. *Journal of Global Positioning System, 2*(1), 42–47.
Kao, J., Flicker, D., Henninger, R., Ghil, M., & Ide, K. (2003). *Using extended Kalman filter for data assimilation and uncertainty quantification in shock-wave dynamics.* Los Alamitos, CA: IEEE Computer Society Press. 3-4942.
Kim, K. H., & Lee, J. G. (2006). Adaptive two-stage EKF for INS-GPS loosely coupled system with unknown fault bias. *Journal of Global Positioning Systems, 5*(1–2), 62–69.

Lee, K. S. (2011). Modeling on the cyclic operation of standing column wells under regional groundwater flow. *Journal of Hydrodynamics, 23*(3), 295–301.

Lin, L., Yang, J., Zhang, B., & Zhu, Y. (2010). A simplified numerical model of 3-D groundwater and solute transport at large scale area. *Journal of Hydrodynamics, 22*(3), 319–328.

Park, E. E., & Zhan, H. H. (2001). Analytical solutions of contaminant transport from finite one, two, and three dimensional sources in a finite thickness aquifer. *Journal of Contaminant Hydrology, 53*(1), 41–61.

Saad, G. A. (2007). Stochastic data assimilation with application to multi-phase flow and health monitoring problems. Doctoral dissertation. Los Angeles, CA: University of Southern California.

Tam, E. K. L., & Beyer, P. H. (2002). Remediation of contaminated lands, a decision methodology for site owners. *Journal of Environmental Management, 64*, 387–400.

Todd, A., Steve, B., Clint, D., Fredrik, S., Chong, W., & Mary, W. (1996). Computational methods for multiphase flow and reactive transport problems arising in subsurface contaminant remediation. *Journal of Computational and Applied Mathematics, 74*, 19–32.

Welch, G., & Bishop, G. (2006). An introduction to the Kalman filter. Technical Report No. TR95-041. University of North Carolina, Department of Computer Science. Retrieved July 20, 2012, from http://www.cs.unc.edu/~welch/media/pdf/Kalman_intro.pdf.

Zheng, C., & Bennett, G. D. (1995). *Applied contaminant transport modeling, theory and practice*. New York: Van Nostrand-Reinhold.

Zou, S., & Parr, A. (1995). Optimal estimation of two-dimensional contaminant transport. *Groundwater, 33*(2), 319–325.

Application of Ensemble Square Root Kalman Filter in a Three-Dimensional Subsurface Contaminant Transport Model

Shoou-Yuh Chang and Torupallab Ghoshal

Abstract A three-dimensional transport model with advection, dispersion, and reaction has been developed to predict transport of a reactive continuous source pollutant. Numerical Forward-Time-Central-Space (FTCS) scheme has been used to solve the advection-dispersion-reaction transport model and Kalman filter (KF) and Ensemble Square Root Kalman Filter (EnSRKF) have been used for data assimilation purpose. EnSRKF uses Monte Carlo simulation in Bayesian implementation to propagate state ensemble without perturbing observation during assimilation period. In this study, contaminant concentration is the state that has been propagated by this model. Reference true solution derived from analytical solution has been used to compare model results. Root Mean Square Error (RSME) profile shows that the EnSRKF concentration estimate can improve prediction accuracy better compared to numerical and KF approaches. With 10,000 mg/L initial concentration numerical scheme shows an average error of 130.84 mg/L, whereas EnSRKF shows an average error of 10.65 mg/L, indicating an improvement of 91.80 %. Kalman Filter (KF) shows an average error of 31.08 mg/L. Therefore, EnSRKF approach reduces mean RMSE by 65.70 % compared to KF approach.

Introduction

Traditional deterministic numerical models are plagued with various limitations to predict transport of contaminant in subsurface environment. They cannot properly handle the uncertain heterogeneity of subsurface environment. Stochastic filtering technique with Bayesian framework is a relatively new approach in hydrology and environmental engineering problems. Of all the Bayesian data assimilation

S.-Y. Chang (✉)
North Carolina A&T State University, Greensboro, NC, USA
e-mail: chang@ncat.edu

T. Ghoshal
Department of Civil Engineering, North Carolina A&T State University, Greensboro, NC 27411, USA
e-mail: tghoshal@aggies.ncat.edu

algorithms Kalman filter (KF) is most popular; however, it works best with linear systems with a Gaussian distribution. Extended Kalman Filter (EKF) can handle nonlinear cases with some extent; however, it cannot handle large systems. Ensemble techniques can handle purely nonlinear large systems quite efficiently. Ensemble techniques use Monte Carlo-based simulation for Bayesian implementation. Huang et al. (2009) used EnKF to calibrate a heterogeneous conductivity field and improve prediction of solute transport with unknown contaminant source in subsurface. Chang and Latif (2010) used EKF in subsurface contaminant transport prediction model with a two-dimensional domain.

In this study, Ensemble Square Root Kalman Filter (EnSRKF), an efficient variant of Ensemble Kalman Filter (EnKF, Evensen 2003) is used for data assimilation purpose. A detailed description of Square Root filters can be found in Tippett et al. (2003). EnSRKF has a wide-ranged use in ocean modeling, numerical weather prediction, etc. Recently, it has been used in hydrology also. Chen et al. (2013) used EnSRKF to assimilate stream flow data in a flood forecasting model.

To conduct this study, a synthetic three-dimensional space-time domain has been developed and a traditional deterministic model is coupled with KF and EnSRKF to evaluate performance of EnSRKF in subsurface contaminant transport. A reactive contaminant with continuous input is modeled with these coupled numerical and stochastic schemes. Performance and effectiveness of EnSRKF is compared with KF and numerical scheme using measurement data simulated from analytical solution.

Methodology

The three-dimensional form of the advection-dispersion-reaction equation for nonconservative pollutant in a saturated, heterogeneous porous media with flow domain in x direction is given by the following partial differential equation:

$$\frac{\partial C}{\partial t} = \frac{D_x}{R}\frac{\partial^2 C}{\partial x^2} + \frac{D_y}{R}\frac{\partial^2 C}{\partial y^2} + \frac{D_z}{R}\frac{\partial^2 C}{\partial z^2} - \frac{V_x}{R}\frac{\partial C}{\partial x} - kC \quad (1)$$

where C = solute concentration, mg/L, t = time, s, D_x, D_y, and D_z = dispersion coefficients in x, y, and z direction, respectively, m^2/day, R = retardation factor, dimensionless; V = velocity, m/day, k = first-order decay rate constant, day^{-1}, x, y, z = Cartesian coordinates, m.

The three-dimensional subsurface transport equation is solved using numerical Forward-Time and Central-Space (FTCS) method. We discretize the partial differential equation in 3D space and the linear system of equation can be written in the following form:

$$\mathbf{x_{t+1}} = \mathbf{M_t(k)x_t} \quad (2)$$

where x_{t+1} is the state variable matrix containing contaminant concentration at time $t+1$, $M_t(k)$ is the state transition matrix containing the degradation decay rate, and k, x_t is the state variable matrix containing contaminant concentration at time t.

Three-dimensional contaminant transport plume can be simulated using assumption of stochastic system with uncertain sources of errors. Process or system equation can be written in the following form:

$$x_{t+1} = M_t(k)x_t + p_t, \quad t = 0, 1, 2, 3, \ldots \ldots \quad (3)$$

where p_t is the process or system error of model. This error is assumed to be a white or uncorrelated noise having a covariance matrix of Q_t. The model error Q_t is considered as a Gaussian distribution with standard deviation of 10 % of concentration. Observation or measurement equation can be written in the following form:

$$z_t = Hx_t^T + O_t, \quad t = 0, 1, 2, 3, \ldots \quad (4)$$

Observation or measurement vector z_t is having simulated measurement value in observation nodes in each time step. Observation data is simulated from true solution derived by solving the advection-dispersion-reaction equation analytically. The equation of Domenico and Schwartz (1990) is used for analytical solution. A 2.5 % random error is added to true solution to obtain the observation data. In this study in total 54 observation nodes are used which are less than 10 % of total 720 nodes. **H** is an operator that denotes observation data pattern. The observation noise vector O_t is chosen from a Gaussian distribution with standard deviation of 2.5 % of concentration. The observation noise vector has covariance matrix R_t.

Kalman Filter (KF). Figure 1 illustrates the sequential operation of Kalman filter to estimate state (Chang and Latif 2007). The whole operation can be divided into two major parts: prediction and correction. With the initial state the filter estimates the Kalman gain. Then, it updates the optimal error covariance matrix before going

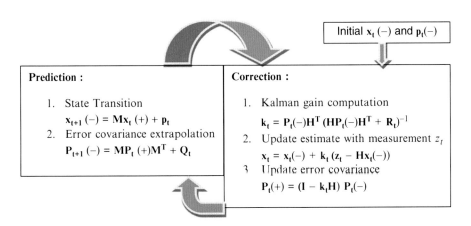

Fig. 1 Sequential operation of Kalman filter

to the prediction step. This is an iterative process and this cycle goes on for estimation of next time step.

Ensemble Square Root Kalman Filter (EnSRKF). EnSRKF is an efficient variant of Ensemble Kalman Filter (EnKF, Evensen 2003). Comparing to EnKF, EnSRKF improves efficiency of analysis by avoiding perturbations of observation during assimilation period (Whitaker and Hamill 2002). EnSRKF also does not have large inversion computation during analysis step which makes it a very efficient tool for data assimilation.

Few steps of EnSRKF implementation are similar to implementation of EnKF. We start the implementation of EnSRKF following the implementation of EnKF described in Chang and Latif (2011). However, after few steps we formulate EnSRKF accordingly. Similar to EnKF, Monte Carlo sampling is performed to generate the ensemble members. N numbers of ensemble members have been generated adding uncorrelated and independent random error. The state matrix is built using the arrangement of ensemble members $x_i \in \Re^n, (i = 1, \ldots, N)$

$$\mathbf{A}^f_{t|t-1} = [\mathbf{x}_1, \mathbf{x}_2, \ldots, \mathbf{x}_N] \in \Re^{n \times N} \tag{5}$$

where n is the size of model state vector. The ensemble mean can be calculated by the following operation.

$$\overline{\mathbf{A}}^f_{t|t-1} = \mathbf{A}^f_{t|t-1} \mathbf{B}, \tag{6}$$

where $\overline{\mathbf{A}}^f_{t|t-1}$ is the ensemble mean matrix and $\mathbf{B} \in \Re^{N \times N}$ is the matrix where each element is equal to $1/N$. The ensemble residual matrix can be defined as

$$\mathbf{E} = \mathbf{A}^f_{t|t-1} - \overline{\mathbf{A}}^f_{t|t-1} \tag{7}$$

Vector of measurements $\mathbf{z} \in \Re^m$, with m being the number of observation grid points, can be stored in the columns of a matrix

$$\mathbf{Z} = [\mathbf{z}_1, \mathbf{z}_2, \ldots, \mathbf{z}_N] \in \Re^{m \times N} \tag{8}$$

Bannister (2012) showed a three-step analysis procedure for EnSRKF. At first step, we find a mean analysis state matrix, $\overline{\mathbf{X}}^a_t$.

$$\overline{\mathbf{X}}^a_t = \overline{\mathbf{A}}^f_{t|t-1} + \mathbf{E}\mathbf{S}^T\mathbf{C}^{-1}\left(\mathbf{Z} - \mathbf{H}\overline{\mathbf{X}}_{t|t-1}\right) \tag{9}$$

where $\mathbf{H}$ is the observation operator; $\mathbf{Z}$ is the observation data matrix and $\mathbf{S}$ and $\mathbf{C}$ matrices are defined below:

$$\mathbf{S} = \mathbf{HE} \in \Re^{m \times N} \tag{10}$$

and $\mathbf{C} = \mathbf{SS}^T + (N-1)\mathbf{R_t} \in \Re^{m \times m}$, where $\mathbf{R_t}$ is the $m \times m$ observation error covariance matrix. Then, we define $\mathbf{G} = \mathbf{S}^T\mathbf{CS}$. We then find the eigenvector $\mathbf{V}$ and eigenvalue $\mathbf{D}$ of matrix $\mathbf{G}$.

In second step, we calculate analysis perturbations, $\mathbf{A}$:

$$\mathbf{A} = \mathbf{EV}(\mathbf{I} - \mathbf{D})^{1/2}\mathbf{V}^T \tag{11}$$

Inversion of $(\mathbf{I} - \mathbf{D})^{1/2}$ involves potential singularity. Therefore, pseudo-inverse has been calculated to avoid this potential singularity.

In final step, we assemble the full ensemble using analysis perturbation and we propagate this ensemble to next time step.

$$\mathbf{A_t^a} = (\overline{\mathbf{X}}_t^a, \overline{\mathbf{X}}_t^a, \ldots, \overline{\mathbf{X}}_t^a) + \mathbf{A} \tag{12}$$

$$\mathbf{A}^f_{t+1|t} = \mathbf{M}(\mathbf{A_t^a}) + \mathbf{p_t} \tag{13}$$

where posterior state matrix at time t, $\mathbf{A_t^a}$ is used to update prior state matrix, $\mathbf{A}^f_{t+1|t}$ at time $t+1$ with a linear state transition operator matrix $\mathbf{M}$. $\mathbf{p_t} \in \Re^{n \times N}$ is a matrix of stochastic error perturbations with an error covariance of $\mathbf{Q_t}$.

Model and Parameters

A three-dimensional volumetric space grid has been created with 10 nodes in X axis, 12 nodes in Y axis, and 6 nodes, i.e., layers in Z axis. In total, it has 720 nodes. Grid spacing Δx, Δy, and Δz in between each node is 3 m. 30 time steps have been used to simulate the transport of concentration and each time step; Δt has duration of 0.75 days. A continuous pollutant source with an injection rate of 10,000 mg/L is inserted in grid point (1, 6, 1). Nine observation nodes has been used in each layer that means in six layers a total of 54 observation nodes has been used in a domain of 720 nodes. The other parameters are chosen according to suggestion by Zou and Parr (1995). These parameters are: velocity 0.8 m/d, porosity 0.3, retardation factor 1.5, hydrodynamic dispersions $D_x = 1.3$ m²/d, $D_y = 0.5$ m²/d, and $D_z = 0.7$ m²/d and first order decay rate is 0.2/d.

Result and Discussion

Several contour profiles are drawn to compare contaminant transport prediction with true solution. To illustrate spread of plumes, two contour profiles have been presented here: one is for time step 10 at layer 2 and another one is for time step 30 at layer 2. In Fig. 2, contaminant contour for time step 10 at layer 2 is shown for True solution, FTCS, KF, and EnSRKF solutions. Concentration plume produced by Numerical FTCS scheme moves faster compared to true solution. FTCS plume also shows a smooth shape which is not congruent with true solution. On the other hand, KF and EnSRKF shows movement pattern closer to true solution and they show irregular shapes similar to true solution plume. However, from scales it is clearly evident that EnSRKF is most successful to produce concentration plume similar to true solution. In Fig. 3, concentration plume distribution of time step 30, i.e., final time step is plotted for layer 2.

This plot also indicates that EnSRKF can better resemble the plume distribution of true solution. In both figures, true solution shows irregular shape of plume distribution which can be attributed to the heterogeneity of subsurface environment and EnSRKF is most successful to adapt this heterogeneity. Numerical scheme showing smooth shaped contour lines cannot represent heterogeneity of subsurface adequately.

To evaluate performance of different approaches, Root Mean Square Error (RMSE) profiles are plotted for numerical FTCS, KF, and EnSRKF approaches using simulated true value as reference solution. The RMSE profiles are shown in Fig. 4. Initially, at the start of the simulation there is no error; therefore, initial error

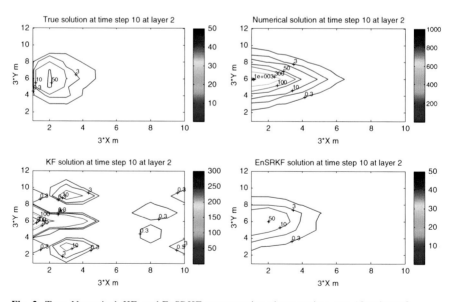

Fig. 2 True, Numerical, KF, and EnSRKF concentration plume at time step 10 at layer 2

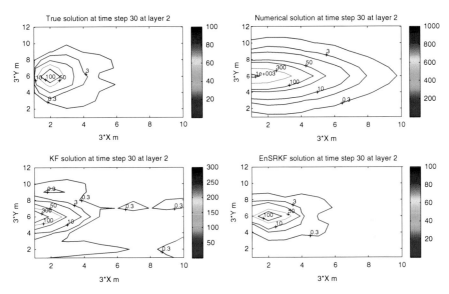

Fig. 3 True, Numerical, KF, and EnSRKF concentration plume at time step 30 at layer 2

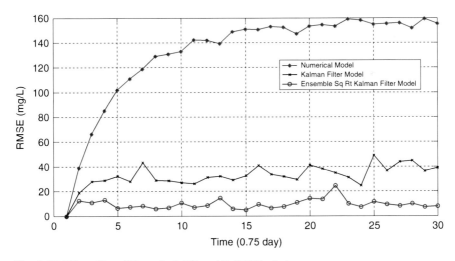

Fig. 4 RMSE profiles of Numerical, KF, and EnSRKF solutions

starts from zero. RMSE profile shows that numerical error is increasing over time for up to 15 time steps, after this time step increment rate of numerical error slows down. KF RMSE shows a relatively steady plot with an error range around 20–30 mg/L. EnSRKF shows least error among these three with less than 20 mg/L for all time steps except only one. All three RMSE profiles show error estimation for a single run. However, as we added random error to prepare simulated true field and due to randomness each run should differ from another. Therefore, to test stability we ran the model for ten times and we calculate average RMSE of each approach. An average RMSE of all time steps is calculated for a single run and an average is calculated taking the averages of ten similar runs. With ten different runs average RMSE for numerical approach is 130.84 mg/L, average RMSE of KF approach is 31.08 mg/L, and that of EnSRKF approach is 10.65 mg/L. Therefore, introduction of EnSRKF has improved average RMSE of 91.8 % compared to numerical scheme and 65.7 % compared to KF approach.

Conclusion

The results obtained from prediction error (RMSE) indicate that true solution is better resembled by EnSRKF compared to numerical FTCS and traditional KF. With less than 10 % observation nodes EnSRKF can improve prediction accuracy by 65.7 % over KF approach and 91.8 % over numerical approach. RMSE profiles show that the error trend of each scheme is not showing significant decreasing pattern as expected from numerical and filtering techniques. This can be explained by the concentration injection consideration used in this test case; we considered a continuous input with 10,000 mg/L concentration injection rate throughout the entire simulation period which produced relatively flatter RMSE trend as a large mass is inserted continuously. However, if we use instantaneous or spill input RMSE trend will go down over time. With significant improvement over numerical and Kalman filter approaches we can conclude that introduction of EnSRKF is quite successful in subsurface contaminant transport model.

References

Bannister, R. N. (2012). A square-root ensemble Kalman filter demonstration with the Lorenz model. University of Reading. Retrieved from http://www.met.reading.ac.uk/~darc/training/lorenz_ensrkf.

Chang, S.-Y., & Latif, S. M. I. (2007). Use of Kalman filter and particle filter in a one dimensional Leachate transport model. In *Proceedings of the 2007 National Conference on Environmental Science and Technology, NC A&T State University*, Greensboro, NC, September 12–14, 2007.

Chang, S.-Y., & Latif, S. M. I. (2010). Extended Kalman filtering to improve the accuracy of a subsurface contaminant transport model. *Journal of Environmental Engineering, 136*(5), 466–474.

Chang, S.-Y., & Latif, S. M. I. (May, 2011). Ensemble Kalman filter to improve the accuracy of a three dimensional flow and transport model with a continuous pollutant source. In *Proceedings of the 2011 World Environmental and Water Resources Congress*, Palm Springs, CA, May 12–13, 2011.

Chen, H., Yang, D., Hong, Y., Gourley, J. J., & Zhang, Y. (2013). Hydrological data assimilation with the Ensemble Square-Root Filter: Use of stream flow observations to update model states for real-time flash flood forecasting. *Advances in Water Resources, 59*, 209–220.

Domenico, P. A., & Schwartz, F. W. (1990). *Physical and chemical hydrogeology*. New York: Wiley-Interscience.

Evensen, G. (2003). The ensemble Kalman filter: Theoretical formulation and practical implementation. *Ocean Dynamics, 53*, 343–367.

Huang, C., Hu, B., Li, X., & Ye, M. (2009). Using data assimilation method to calibrate a heterogeneous conductivity field and improve solute transport with an unknown contaminant source. *Stochastic Environmental Research and Risk Assessment, 23*(8), 1155–1167.

Tippett, M. K., Anderson, J. L., Bishop, C. H., Hamill, T. M., & Whitaker, J. S. (2003). Ensemble square-root filters. *Monthly Weather Review, 131*, 1485–1490.

Whitaker, J. S., & Hamill, T. M. (2002). Ensemble data assimilation without perturbed observations. *Monthly Weather Review, 130*, 1913–24.

Zou, S., & Parr, A. (1995). Optimal estimation of two-dimensional contaminant transport. *Ground Water, 33*(2), 319–325.

Application of 3D VAR Kalman Filter in a Three-Dimensional Subsurface Contaminant Transport Model for a Continuous Pollutant Source

Shoou-Yuh Chang and Anup Saha

Abstract Modeling of contaminant transport in a subsurface environment by a deterministic model deviates from the real world environment with time because of the highly heterogeneous nature of the subsurface environment. In this study, an optimized stochastic approach coupled with the deterministic model has been applied to predict the contaminant transport in a subsurface environment. A three-dimensional contaminant transport model for a continuous pollutant source has been developed for this purpose. A forward time center space (FTCS) model has been used as a numerical approach to solve the classical advection dispersion flow equation and the Kalman Filter (KF) coupled with 3D variational analysis has been used for data assimilation purpose. The Kalman Filter is a recursive-based algorithm consists of a set of mathematical equations, which takes a prior estimation and observation measurements into consideration to compute the best prediction for a state variable. The 3D variational analysis uses a cost function to find out an optimal solution for an analysis using the KF simulated state as its background. The analysis for 3D VAR is found iteratively minimizing the cost function. A Root Mean Square (RMSE) and Mean Absolute Error (MAE) profile were used to evaluate the efficiency of the analysis. The investigation shows that state prediction to be good for both the background (KF) and analysis (3D VAR KF) steps compared to the deterministic solution whereas the 3D VAR analysis over the background (KF) found to be effective in reduction of the error significantly. An overall improvement of 15.3 % error reduction on RMSE and 14.6 % reduction on MAE over the background is achieved using this new approach.

S.-Y. Chang (✉)
North Carolina A&T State University, Greensboro, NC, USA
e-mail: chang@ncat.edu

A. Saha
Department of Civil Engineering, North Carolina A&T State University, Greensboro, NC 27411, USA
e-mail: asaha@aggies.ncat.edu

Introduction

Contamination of ground water is one of the major environmental issues nowadays. The ground water flow and the transport of pollutant to the ground water can be considered as a practical problem. As the soil formations vary rapidly, it is very difficult to get the exact information about the subsurface reservoir system and how the pollutant would behave in the subsurface system. There are a lot of uncertainties involved in solving the fate and pollutant transport problems. So the results in a fate and pollutant model may become erroneous due to the randomness of the system and also if the assumptions made are inaccurate. Thus, it has become a challenge for the engineers to accurately predict the pollutant transport into the subsurface environment. Therefore, the analytic mathematical approach does not always predict the real picture of the situation as it takes the average data and a few number of system parameters into considerations. And also due to the inherent randomness of the system, probabilistic variability of the system, model parameters variation, truncation, or round off error and noisy system, it becomes very difficult for the analytic or deterministic model to predict the real behavior of the pollutant flow into the subsurface environment. Moreover, the pollutant prediction in ground water and the transport of the pollutant in the subsurface system depends on the widely varying information of the heterogenic aquifer system, reaction, and decay mechanism of the system and advection and dispersion mechanism of the system. Thus, it is impossible to attain a perfect deterministic ground water flow and transport model, as even a little error will grow incorrigibly with time. Filtering techniques when combined with traditional deterministic approaches can give us cost-effective results where constructions of observation wells are very expensive. The Kalman Filter essentially consists of a set of mathematical equations, which takes a prior estimation and observation measurements into consideration to compute the best prediction for a state variable. By calculating the covariances of state and observations, these equations are solved. The limitation of Kalman Filter is that it will give optimal measure if the system is linear. Chang and Latif (2009) used Kalman filtering and particle filtering to model the transport of benzene leachate from industrial landfill to the subsurface where Kalman filter found to give up to 80 % less error in comparison to conventional numerical approach. Several studies have been made using these filtering techniques for data assimilation to improve model prediction efficiency. Geer (1982) used Kalman filters to simulate one-dimensional phreatic groundwater flow where he showed a relationship between interpolation confidence of ground water levels to the observation wells distance. Zou and Parr (1995) used the state-space estimation technique with Kalman filtering to estimate optimal concentration in a 2D plume in subsurface environment. Hamill and Snyder (2000) demonstrated a hybrid approach of Ensemble Kalman Filtering with 3D variational analysis using the quasigeostrophic model for perfect-model assumptions. They replaced the background-error covariance matrix and used a weighted sum of the 3D VAR background-error covariance. It was found that the system gave good results for large ensemble size and when the background-error covariances were calculated from ensemble members. Thornhill et al. (2012) modeled the hydrodynamic and sediment transport of Morecambe Bay, UK using the integration of a Morpho-dynamic model with the 3D variational data assimilation technique to assess the prediction efficiency of the newly coupled model in comparison with the model stand alone. The calculated mean square error (MSE) and brier score showed a significant improvement using the data assimilation technique with 3D variational analysis over the model alone performance.

Methodology

Three-dimensional contaminant transport model. The advection-dispersion equation with flow dominant in x-direction and in a three-dimensional heterogeneous saturated porous media for reactive pollutants given by Schwartz and Zhang (2004) is used for contaminant transport model in the subsurface environment.

$$R\frac{\partial C}{\partial t} = D_x\frac{\partial^2 C}{\partial x^2} + D_y\frac{\partial^2 C}{\partial y^2} + D_z\frac{\partial^2 C}{\partial z^2} - v(x,y,z)\frac{\partial C}{\partial x} - \kappa RC, \qquad (1)$$

where R = retardation factor; C = is the solute concentration, mg L^{-1}; t = time, day; D_x, D_y, and D_z = hydrodynamic dispersions in the x, y, and z directions, respectively, m^{2}day^{-1}; $v(x, y, z)$ = linear pore water velocity at the point (x, y, z), mday^{-1}; and $\mathbf{k}$ = is decay constant of pollutants, day^{-1}.

The boundary conditions for the 3D solute transport with a continuous pollutant source is given as

$$C(x_o, y_o, z_o, t)_{t=0} = C_o; \frac{\partial c}{\partial x} = \frac{\partial c}{\partial y} = \frac{\partial c}{\partial z} = 0, \text{ for } x = y = z = \infty$$

Here, (x_o, y_o, z_o) is the pollutant input point and C_o is the continuous input pollutant concentration, mg L^{-1}. A forward time central space (FTCS) finite difference approach is used to solve the transport equation deterministically. After time-space discretization in the 3D space, the linear system of equations can be rewritten in the following vector–matrix form:

$$x_{t+1} = \mathbf{A}_t x_t, \qquad (2)$$

where x_{t+1} and x_t are the solute concentration at time step $t+1$ and t, respectively, $\mathbf{A}_t$ is the linear operator or state transition matrix that predicts the next time step state vector from the current state vector.

The three-dimensional contaminant transport model can be simulated as stochastic which can be represented by the equation

$$x_{t+1} = \mathbf{A}_t x_t + P_t, \ t = 0, 1, 2, 3, 4 \ldots \qquad (3)$$

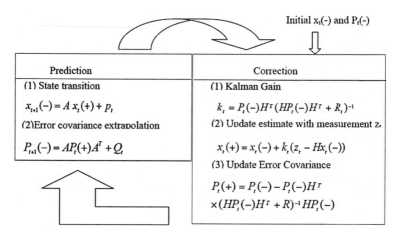

Fig. 1 Discrete Kalman filter operation

where P_t is the process noise which is assumed to be normally distributed having zero mean and a standard deviation of 5 %.

Similarly, we can also simulate the observation equation as follows:

$$z_t = H_t x_t^T + O_t, \ t = 0, 1, 2, 3, 4 \ldots \ldots \tag{4}$$

where z_t is the observation value of state of the nodes at time t, x_t^T is the true value of a state of any node, H_t is the observation data pattern matrix, and O_t is the observation error which is also taken from a Gaussian distribution with standard deviation of 2.5 %. Figure 1 shows the recursive operation of the discrete Kalman filter process.

The Kalman filter predicted state then used as background state for the 3D VAR analysis. The basis of 3D-VAR methods is to minimize the objective function by obtaining an optimum solution of x such that $x = x_a$. The function is written as (Hamill and Snyder 2000),

$$J(x) = \frac{1}{2}\left[(x - x_b)^T B^{-1}(x - x_b) + (y - Hx)^T R^{-1}(y - Hx)\right] \tag{5}$$

where x is a model state vector, x_b is the background state vector, y is the observation vector, H is the linearized observation operator that maps model variables to observation space, B is the covariance matrix of background errors, and R is the covariance matrix of observation errors.

The analysis increment $(x_a - x_b)$ satisfies the following equation:

$$\begin{aligned}(I + BH^T R^{-1} H)(x_a - x_b) &= BH^T R^{-1}(y - Hx_b) \\ \Rightarrow x_a &= x_b + (I + BH^T R^{-1} H)^{-1} BH^T R^{-1}(y - Hx_b)\end{aligned} \tag{6}$$

where x_a is the analysis state vector; I is the identity matrix.

Equation (6) cannot be solved directly because of the huge inversion of matrix. The analysis is rather achieved by taking the gradient of the (5) and solving it iteratively. A model grid of 10 × 10 × 1 with 2 m spacing was used to simulate the model with total 30 time steps; each time step of 0.75 days. The parameters used in this model are ground water velocity = 0.5 m/d, porosity = 0.3, dispersions $D_x = 1$ m²/d, $D_y = 0.8$ m²/d, $D_z = 0.5$ m²/d in x, y, and z direction, respectively, decay rate $(k) = 0.1$/d, continuous pollutant input = 10,000 mg/L, retardation factor, $R = 1.5$, initial pollutant input point, $x_0 = 1$, $y_0 = 5$, $z_0 = 1$.

Results and Discussion

The simulation for contaminant transport after 3.75 days and 21.5 days has been showed (Figs. 2 and 3) for three different models used. It can be seen that both the stochastic model Kalman filter and 3D VAR Kalman filter gave better prediction of

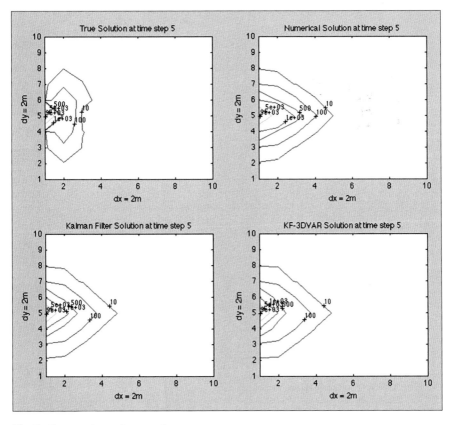

Fig. 2 Contour plot at time step 5

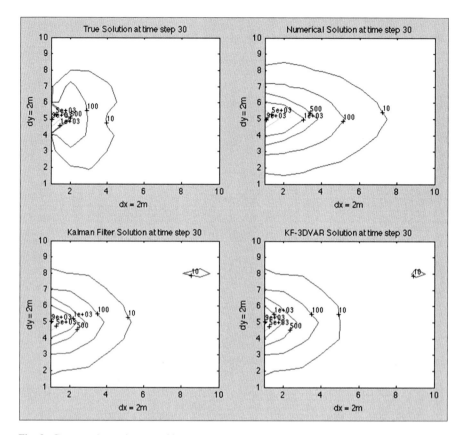

Fig. 3 Contour plot at time step 30

the contaminant state compared to the FTCS model and the FTCS model tends to move faster compared to the other two models.

Root mean square error (Fig. 4) in the FTCS model tends to increase with time resulting from the inclusion of contaminant at each time step making the FTCS model a poor convergence. On the other hand, the other two models also seemed to give higher error at starting of the simulation but tended to converge as time increases. Although from Figs. 3 and 4, it seems that both the KF and 3D VAR KF show very close prediction for both the stochastic approaches but from the Fig. 5 we can find that there is a significant error improvement over the background (KF) by the analysis (3D VAR KF). The mean of the root mean square error (RMSE) is 324.8 mg/L, 168.2 mg/L, and 142.4 mg/L for FTCS, KF, and 3D VAR KF, respectively. The Kalman filter reduced error up to 48.2 % whereas 3D VAR KF reduced error up to 56.2 % compared to the numerical model. Thus, the new approach gave an overall 15.3 % improvement over the background (KF).

The mean absolute error (MAE) profile in Fig. 5 shows that the error profile for both the stochastic approaches very less compared to the deterministic approach

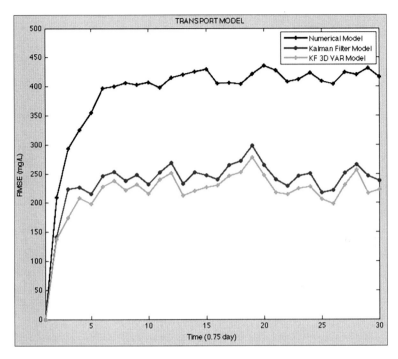

Fig. 4 RMSE profile for Numerical Model, KF and 3D VAR KF

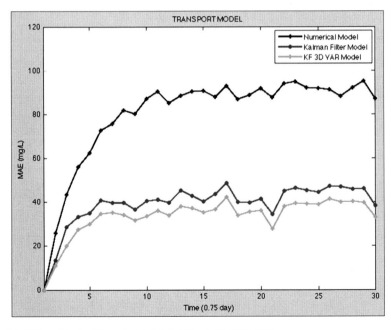

Fig. 5 MAE profile for Numerical Model, KF and 3D VAR KF

giving a mean absolute error of 79.8 mg/L, 39.1 mg/L, and 33.4 mg/L for FTCS, Kalman filter, and 3D VAR Kalman filter approaches, respectively. 3D VAR analysis on the background Kalman filter prediction gave an overall error improvement of 14.6 % in MAE profile.

Conclusion

The accuracy of the stochastic ground water contaminant transport model greatly depends on the proper inclusion of the state variables and model parameters. Due to the high heterogeneity of the subsurface, it is impossible to introduce all the parameters accurately. In context to that, the stochastic models used in this study found to be quite efficient. Although Kalman filter works well for the linear problems, the state prediction might not be able to achieve desired level of accuracy. But still compared to the deterministic model it is quite efficient. And as the accuracy of the new approach 3D VAR KF depends on its background state prediction; the more accurate the background the more accurate the analysis would be. Being the iterative process, 3D VAR analysis could be computationally expensive. But it is still worth where prediction accuracy of a contaminant is very vital with a little extra computational cost. Nonlinear models such as Ensemble Kalman filters could be used successfully with the 3D variational analysis to achieve more prediction accuracy and fewer errors.

References

Chang, S.-Y., & Latif, S. M. I. (2009). Use of Kalman filter and particle filter in a one dimensional Leachate transport model. In *Proceedings of the 2007 National Conference on Environmental Science and Technology* (pp. 157–163). New York: Springer.

Geer, F. C. V. (1982). An equation based theoretical approach to network design for groundwater levels using Kalman filters. *International Association of Hydrological Sciences, 136*, 241–250.

Hamill, T. M., & Snyder, C. (2000). A hybrid ensemble Kalman filter—3D variational analysis scheme. *Monthly Weather Review, 128*(8), 2905–2919.

Schwartz, F. W., & Zhang, H. (2004). *Fundamentals of ground water*. New York: Wiley.

Thornhill, G. D., Mason, D. C., Dance, S. L., Lawless, A. S., Nichols, N. K., & Forbes, H. R. (2012). Integration of a 3D variational data assimilation scheme with a coastal area morphodynamic model of Morecambe Bay. *Coastal Engineering, 69*, 82–96.

Zou, S., & Parr, A. (1995). Optimal estimation of two-dimensional contaminant transport. *Ground Water, 33*(2), 319–325.

Groundwater Flow Modeling in the Shallow Aquifer of Buffalo Creek, Greensboro

Jenberu Feyyisa, Manoj K. Jha, and Shoou-Yuh Chang

Abstract U.S. Environmental Protection Agency categorized Buffalo Creek as one of the impaired waters in Cape Fear River Basin, North Carolina, due to high concentrations of metals and pathogens. These contaminants originate from effluents discharged from industries and agricultural activities. This study used a numerical groundwater flow modeling approach to investigate the surface and groundwater interaction in the Buffalo Creek watershed for the fate and transport of contaminants. The movement of groundwater flow was simulated using MODFLOW while the particle tracking was analyzed by the MT3D model. MODFLOW solves groundwater flow equation using the finite-difference approximation. The flow region which covers both North and South Buffalo Creek was subdivided into blocks or cells in which the medium properties were assumed to be uniform. The cells are made from a grid of mutually perpendicular lines that are variably spaced depending upon the location. Spatial locations and distributions of stream networks, elevations, boundary conditions (no flow and constant, variable head zones), existing well locations, and industrial as well as wastewater effluent discharge locations were developed within ArcGIS environment. The modeling tasks of this study are domain characterization (database for surface elevation and stream network), modeling setup, and calibration and validation of the model using observed data. The observed data for baseflow was obtained using the baseflow filter algorithm, which basically separates baseflow from streamflow based on nature of the hydrograph. The modeling setup and initial calibration results for the steady-state simulation are presented in this chapter.

J. Feyyisa • M.K. Jha (✉)
Department of Civil, Architectural, and Environmental Engineering, North Carolina Agricultural and Technical State University, Greensboro, NC, USA
e-mail: mkjha@ncat.edu

S.-Y. Chang
North Carolina A&T State University, Greensboro, NC, USA
e-mail: chang@ncat.edu

Introduction

Due to its more complex nature, the movement of particles and their effects in groundwater system cannot be traced and detected easily. Because of this, much of resources required to study and mitigation measures are skewed towards other water resources (surface water) sectors. Timeframes between an original pollution event, percolation through the unsaturated zone, transport in groundwater, and eventual baseflow discharge to a receiving river may be years to decades and depend upon the pathways and distances involved, groundwater velocities and capacity for natural attenuation of a pollutant in the subsurface. On the contrary, the effect of contaminated surface waters (streams, rivers, lakes and ponds) can easily be detected through their direct symptoms like aquatic life deterioration, human health changes and others, attracting the attention of experts, managers and politicians. The development of human activities expanded in cities, agricultural lands and in large settlements. Groundwater contamination by number of metallic elements and organics is nearly always the result of human activities such as hazards which pose health risks to human.

In almost all of the previous studies and assessments, much attention was given only to surface waters where immediate effects can be observed. Due to the critical role of groundwater in the hydrologic cycle and ecosystem, funding decisions to prevent adverse effects to the resources will more fully recognize groundwater's role. Unfortunately, the complex nature of groundwater movement and its interaction with surfaces waters (lakes, streams, rivers and recharge) may be one of the reasons why much was not done in this area. The amount of groundwater available from the regolith-fractured crystalline rock aquifers system in Guilford county, North Carolina, is largely unknown. To begin to address pollution prevention or remediation, we must understand how groundwater and surface water interrelate. These are interconnected and can be fully understood and intelligently managed if only when the fact is acknowledged. The need to understand groundwater system and its interaction with surface water and stressors (contaminants) is one of the areas where recent studies give attention due to its paramount need. Planners and Managers benefit from additional knowledge of groundwater resources.

This study attempts to build a calibrated steady-state flow of groundwater in the regolith aquifer of Buffalo Creek Watershed, Guilford County. A three-dimensional steady-state model MODFLOW (PMWIN 5.3.1) was constructed to simulate shallow aquifer of groundwater flow. ArcGIS was used for geographic data development, management, integration, and analysis. Using tools of ArcGIS such as ArcMap, Arc Catalogue, and ArcTools, surface elevations, streamflow networks, flow directions, flow lengths, basin boundaries, and slopes are created, analyzed, and used for MODFLOW inputs. Based upon the limitations of PMWIN to process the input, ArcGIS tool provides flexibility in setting up the appropriate spatial allocations and merging of the default raster data through resampling and extraction of DEM data. MODFLOW is used to numerically solve the three-dimensional groundwater flow equation using the finite-difference method (Harbaugh et al. 2000). The groundwater flow equation is solved using the finite-difference approximation method. The flow region is considered to be subdivided into blocks in which the medium properties are assumed to be uniform. The plan view rectangular discretization results from a grid of mutually perpendicular lines that may be variably spaced.

Site Description

This study was conducted on 88.5 mile^2 of Buffalo Creek watershed in Guilford County, North Carolina. Guilford County consists of approximately 658 mile^2 in the central part of the Piedmont province. The watershed is located in the upper Cape Fear River basin that includes most parts of Greensboro city, characterized by shallow unconfined regolith aquifer. The topography of the area consists of low rounded hill and long, northeast-southwest trending ridges with up to few hundred feet of local relief. The Climate of Guilford County is typed as humid-subtropical with mean minimum January temperatures range from 31 to 33 °F whereas mean maximum July temperature range from 87 to 89 °F. Annual precipitation varies across the county from 43 to 48 in. (Kopec and Clay 1975). The lowest rainfall occurs in the southern and south western parts of the county; the highest rainfall occurs in the southern and southeastern parts of the county (HERA Team 2007).

Geologically, Guilford County lies within the Charlotte Slate Belt. Metamorphic and Igneous crystalline rocks are mantled by varying thickness of regolith (HERA Team 2007). Daniel and Harned (1998) in their idealized sketch of groundwater system, categorized the regolith aquifer geological setting of the Guilford County as: (1) the unsaturated zone in the regolith, which generally contains the organic layer of the surface soil, (2) the saturated zone in the regolith, (3) the lower regolith which contains the transition zone between saprolite and bedrock, and (4) the fractured crystalline bed rock system.

Methodology

ArcGIS 10.0 tools (ArcMap, ArcCatalogue, and ArcTools) were used to manage and transform spatial distribution of geospatial parameters. First, from a USGS developed earth explorer interface, data of Shuttle Radar Topographic Mission (SRTM) 2000, an approximated geospatial area of the watershed is sited and delineated to be downloaded as a rectangular area (79–80° W and 35–36° N). The approximated region of interest was tiled into two subsets for downloading and later exporting to GIS.

In ArcGIS, the two subset data tiles were imported as mosaic data and retiled as a single data using data management tools. Depending upon the PMWIN limitations in data inputs (maximum cell size of 250,000), raster data of DTED resampled

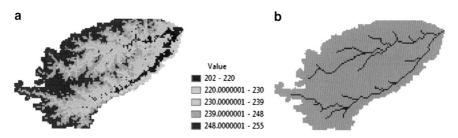

Fig. 1 (a) Delineated Buffalo Creek Watershed and (b) stream network and distribution

multiple times until maximum allowed or less number of cell sizes and appropriate cell dimensions are met. After filling the raster data set for any sink, a cell with defined drainage direction, flow directions were developed for major and tributary streams using hydrology tools. While establishing the stream network, output cells with high flow accumulation are only used to identify streams. For the size of cells we have already set, high flow accumulation is considered for cells receiving flow from more than 100 output cells. That means cells with flow accumulation receiving from zero output raster are categorized as highs or peaks that also indicates the boundary of the watershed. Consecutively, cells receiving flow from 0 up to 99 numbers are categorized as undefined direction flow cells. The final feature of the stream channel and links are constructed by thresholding the results of the flow accumulated raster and using GIS conversion tools (Fig. 1b). While delineating the targeted basin, the analysis extent is narrowed and widened from the original coordinate until the outlet from the basin in question is delineated and visible (Fig. 1a).

PMWIN is one of the most complete groundwater simulation systems in the world developed by Chiang and Kinzelbach (2001). It includes all the supporting models (MODFLOW, MT3D, MT3DMS, MOC3D, PMPATH) for Windows, PEST2000, and UCODE. MODFLOW is a simulation system for modeling groundwater flow with the modular three-dimensional finite-difference groundwater model developed by the U.S. Geological Survey (McDonald and Harbaugh 1988). At present, PMWIN supports seven additional packages which are integrated in the original MODFLOW. One of these packages is the streamflow-Routing package (STR1). This particular package (Prudic 1989) is designed to account for the amount of flow in streams and to simulate the interaction between surface streams and groundwater.

MODFLOW is a computer program that simulates one-, two-, or three-dimensional groundwater flow using a finite-difference solution of the model formulation. The partial differential equation for transient three-dimensional groundwater flow in heterogeneous and anisotropic medium, for confined or unconfined aquifer is expressed as:

$$\frac{\partial}{\partial x}\left(K_{xx}\frac{\partial h}{\partial x}\right) + \frac{\partial}{\partial y}\left(K_{yy}\frac{\partial h}{\partial y}\right) + \frac{\partial}{\partial z}\left(K_{zz}\frac{\partial h}{\partial z}\right) - W = S_s\frac{\partial h}{\partial t} \qquad (1)$$

where K_{xx}, K_{yy}, and K_{zz} are the hydraulic conductivity along the x, y, and z coordinate axes parallel to the major axes of hydraulic conductivities; h is the potentiometric head; W is the volumetric flux per unit volume representing source; S_s is the specific storage of the porous medium; and t is time. MODFLOW is designated to simulate groundwater flow system in the aquifers in which (a) saturated flow condition exists, (b) Darcy's Law applies, (c) the density of groundwater is constant, and (d) the principal directions of horizontal hydraulic conductivity or transmissivity do not vary within the system. The groundwater flow equation is solved using the finite-difference approximation. The flow region is subdivided into blocks in which the medium of properties are assumed to be uniform.

For Buffalo Creek watershed, a two-dimensional steady-state MODFLOW model was constructed to simulate groundwater flow for the shallow regolith unconfined aquifer. The model requires the use of Streamflow routing package and Recharge package. The discretization package also sets the spatial and temporal dimensions. In the basic packages boundary conditions (weather flow in the cell is constant, variable/calculated or no flow) and initial hydraulic heads values defined. Values of hydraulic conductivity and the wetting capability are defined in the block-centered flow package. For all MODFLOW/PMWIN packages, an ASCII input was defined with strict format. From ArcGIS output we finally selected a horizontal grid dimension of 200 m × 200 m size organized along rows and columns. The total grid is divided into 128 columns and 82 rows.

Results and Discussion

Groundwater flow system needs to be simulated and calibrated before determining the fate of contaminants in the regolith aquifer system. Groundwater flow has been simulated and calibrated for steady-state case of the aquifer. The trial and error modeling effort for assumed horizontal hydraulic conductivity, evapotranspiration, recharge flux, and initial hydraulic heads were not succeeded. Parameter values continue to be very sensitive, particularly for horizontal hydraulic conductivity and evapotranspiration rates without giving some convergence trend neither to the volumetric water budget balance nor to the representative hydraulic head. For any change to the order of 10^{-14}, the net volumetric water budget (inflow–outflow) changes to nearly 62 %. Similarly, steady-state head also changes with change in horizontal hydraulic conductivity. For any slight changes like, 10^{-08} in hydraulic conductivity value, calculated hydraulic head values goes to the extent where, in some cells beyond the land surface while in others the cells dry out. PEST, parameter estimation tool included in PMWIN which in fact runs outside of the

MODFLOW PMWIN domain was also tried to calculate model parameters. This inverse modeling approach configured to estimate horizontal hydraulic conductivity, evapotranspiration, and recharge rate. Unfortunately, none of the monitoring wells those intended to provide measurement data for inverse modeling were found in the Buffalo Watershed. The nearest available wells from neighboring watersheds are: Gibsonville monitoring well (G 50W2), an active monitoring well located at the real-world coordinates 36.088262 (x) and 79.547915 (y) located at about 10 km away from the exit point of Buffalo watershed. The other monitoring well, inactive well is Yow 2, is located at about 5 km away from the southern water divide line at coordinates 35.958388 and 79.838645, x and y, respectively. Using the parameter estimation tool, PEST provided in the PWMIN, the coordinates were changed systematically to be located within the watershed boundary region, in such a way that their locations are transformed to the nearest coordinate and also similar ground surface elevation and/or assumed hydraulic head. This approach also fails without arriving at some result to model the aquifer parameters. We conclude that, PEST cannot run (inverse run) for observation wells not only located outside the watershed boundary region, it also not considers even if these wells are not active.

Back to the trial and error procedure, we decided to reduce (with justifiable reason) the number of parameters for estimation. Accordingly, the previous recharge flux that includes the component of evapotranspiration is removed from the input data. While doing so, the annual net groundwater recharge amount is estimated from previous study sources (Daniel and Harned 1998). For this particular case, the seasonal variations of groundwater depth in Guilford County and adjacent Counties were also reviewed to estimate the annual recharge flux for the watershed. The annual average amount of recharge flux can be estimated from the annual average baseflow that groundwater contributes to the stream. It is estimated that baseflow from the catchment has no any other source than areal recharge, which is the percolated part of rainfall.

Baseflow Filter Program was used to separate the annual average amount of discharge from groundwater. The model separates the baseflow from its direct input, streamflow records. This recursive digital filter method described by Nathan and McMahon (1990) was originally used in signal analysis and processing (Lyne and Hollick 1979). Filtering surface runoff (high frequency signal) from baseflow (low frequency signals) is analogous to the filtering of high frequency signals in signal analysis and processing (Arnold and Allen 1999). The stream record data passed over the filter three times (forward, backward, and again forward). Each pass will result in less baseflow as a percentage of total flow. Accordingly, the user gets some added flexibility to adjust the separation to more approximate site conditions. The equation for the filter is:

$$q_t = \beta \ q_{t-1} + (1+\beta)/2 \ast (Q_t - Q_{t-1}) \qquad (2)$$

where q_t is the filtered surface runoff (quick response) at the t time step, Q_t is the original streamflow, and β is the filter parameter. Baseflow, b_t is calculated with the equation

$$b_t = Q_t - q_t \tag{3}$$

From USGS daily stream discharge data (1998–2013) for Buffalo Creek stream, records were downloaded and exported to the Filter program. The filter program outputs the calculated baseflow for the three round run in addition to the measured input discharge. From the separated baseflow of 15 years discharge measurement for Buffalo Creek Watershed at Oak Ridge, the average baseflow is calculated as 55 ft^3/s. This value includes the discharge released to the stream from the two sewage treatment plants: T.Z. Osborne Plant and the North Buffalo Facility. The treatment plants permitted to process about 40 million gallons and 16 million gallons of sewage per day for both T.Z. Osborne and North Buffalo, respectively. The discharge estimated from the plant is assumed to be nearly equal to the total water supply required/supplied for Greensboro City. Nearly, all the sewage or wastewater that is generated by customers flows by gravity through sewers that range from 6 to 72 in. in diameter (City of Greensboro 2011). The total daily treated sewage released to Buffalo Stream is estimated to the average collected sewage from the city. Every day, an average of 26 million gallons of sewage generated in our homes and industries that must be collected, transported, and treated to very stringent standards before it is released back into our environment, our streams. The daily demand of water supply for Greensboro City grows from 12 million gallons in 1960 to 33.4 million gallons in 2013. Considering the average sewage disposed to the stream, it is assumed that a 27 millions of gallons of treated sewage is released to Buffalo Stream daily, the amount that matches closely with the city's daily water supply demand. Finally, this amount is subtracted from the calculated total baseflow from the stream to arrive at an estimated actual baseflow from Buffalo Creek with a discharge rate of 13.23 cubic feet per second. This amount when compared to report by City of Greensboro looks bit higher. The City in its report of Sewage Collection and Reclamation Plant Report for 2011 mentioned that our discharge flow constitutes over 97 % of the stream below our discharge points at the lowest streamflows. This report did not mention how and which time of the month or year this estimation was done, since there are large discrepancies of flow happens over seasons of the year. Moreover, the comparison did not mention

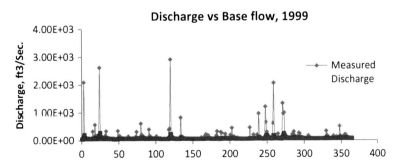

Fig. 2 Streamflow vs. baseflow, typical 1999 yearly average

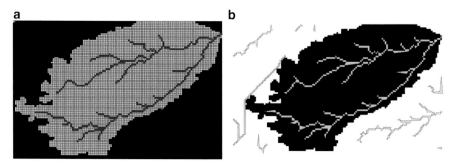

Fig. 3 Geometry of stream network developed by (**a**) MODFLOW-PMWIN and (**b**) GIS

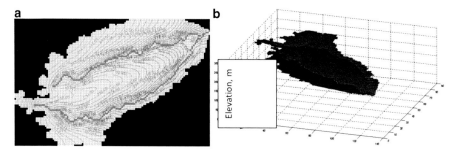

Fig. 4 Steady-state simulated hydraulic heads (**a**) contour and (**b**) 3-D view

weather it is daily, monthly, or annual based or certain period's volumetric estimation. In other words, the assumption/measurement may also be during peak discharging time. Both the true baseflow and discharge from the treatment plants vary significantly over months of the year, but the report does not support its assumption in neither of the cases. In our case, the percentage of actual estimated baseflow to total baseflow (including sewage from treatment plant) estimated to be 21 % (Fig. 2).

To estimate the areal recharge for MODFLOW input, the total volume of the true estimated baseflow amount is distributed over the watershed area. An estimated areal recharge flux of 2×10^{-09} m/s is used as an input recharge rate for the model as an initial input value. After all the required data and parameters input (basic package, discretization package, streamflow routing package, block-centered flow packages, recharge package, porosity) to MODFLOW-PMWIN, the stream network geometry is compared with the natural trend of stream network developed from ArcGIS (Fig. 3).

PMWIN has some limitations on the number of stream networks. The maximum allowed stream segment is twenty-five, and the total number of tributary streams to a segment of stream is ten. Taking into account this limitation, only two small tributary streams on North Buffalo Creek, near where it joins South Buffalo Creek have been omitted. In regard to the number of tributary stream to a segment of stream, we have had only a maximum of three tributary streams that release to a segment, which is less than ten (Fig. 4).

Accounting for the large amount of flow coming from the sewage treatment plant in PMWIN-MODFLOW environment was one of the challenges we faced to run the MODFLOW in arriving at true/representative parameters. Different approaches: (a) distributing the total annual flow as a recharge flux, (b) considering the equivalent discharge as recharging well, by activating well package, and (c) creating an arbitrary river channel by activating river package with equivalent discharge rate as if the river is crossing Buffalo watershed. For the first case, even though the model was able to run properly, the results obtained were at the expense of horizontal hydraulic conductivity values. Conductivity values were increased to the order of 10^2, which are not representative of the watershed's hydraulic conductivity values. For the second case, we tried to recharge the aquifer at the bottom end of one of the top layer cells, but this still leads to accept the recharge as groundwater and influences parameters of the aquifer, nonrepresentative result. Similarly, the river package input was also unable to give a representative water budget values. Fortunately, the last option (option iv) that assumes the treatment plant's discharge as a specified constant head pool in the main stream channels, over the modeling period was able to model groundwater flow effectively. Accordingly, the two main stream channels (North and South Buffalo streams) were considered as constant head cells with varying depth throughout their length.

The distributions of the calibrated hydraulic heads are compared and verified with the available information and previous studies. Within the water shed, a study conducted on Functional Assessment for a Proposed Floodplain Storm Water Treatment Wetland (Kimberly 2002) indicates that groundwater fluctuates between 0.5 and 3 m below ground surface. Even though this study was limited to specific area of the watershed (in regard to groundwater study), it highlights the depth at which groundwater is found in flood plain of South Buffalo sub-watershed. The simulated result of hydraulic heads in flood plain region varies from some few centimeters above ground surface to 5 m below. The variation is acknowledged as the study of groundwater was only confined to an area of one cell width (200 m) of MODFLOW. Daniel and Harned (1998) also mentioned that the seasonal variation of depth of saturated aquifer in piedmont region varies between 4 and 12 ft. They also validated their assumptions from study conducted by on Groundwater Observation Wells in Orange County that this depth of aquifer variation is within 42–46 ft. below the ground surface. The calculated hydraulic heads in our case, although falls within this range, shallower and deeper groundwater depths are also observed. Basically, water wells are selected and located in areas where groundwater can easily be found (technical, economical, and natural reasons). Similarly, it is obvious to assume the locations of these observation wells in relatively less elevated locations indicating that in relatively high elevated nearby locations, the depth of ground water also increases from its reference, ground surface.

Groundwater depth data of relatively nearby observation wells found in adjacent watersheds Gibsonville and Yow 2 were also considered as one of validating parameters. Yow 2 is about 5.3 km south of Buffalo Watershed boundary.

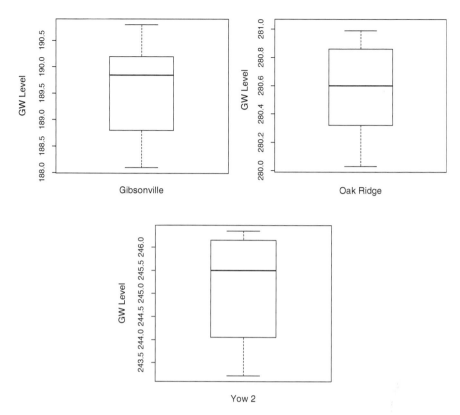

Fig. 5 Box plot, annual groundwater depth fluctuations statistics for Gibsonville, Oak Ridge, and Yow 2 Observation wells

Watershed divide line passes between Yow 2 observation well and similar elevation surface within Buffalo Creek watershed. The distance of this comparison point is nearly 5 km from its watershed divide line. This shows a similar surface slope in either side of the watershed divide line. Gibsonville, the second nearby and only active well located at about 10 km downstream of Buffalo Creek watershed. Available groundwater depth data for Yow 2 (2001–2003) and Gibsonville (2000–2013) is downloaded from USGS/NCWRD website for statistical analysis. Both wells are located at ground surface elevation of 246.88 and 197.5 m for Yow 2 and Gibsonville, respectively. Annual average of groundwater depth is estimated from Box plot of Fig. 5. The similarity of simulated hydraulic head for Yow 2 has been discussed above.

For Gibsonville, the average maximum and minimum data were evaluated and its average annual groundwater depth was estimated to be 189.5731 m. As the topography of the surface between the exit point of Buffalo Creek and the observation well is flat, a constant slope is estimated from distance and altitude, 0.055 %. The gradient of hydraulic head is also assumed to be parallel to ground surface slope. Calculated hydraulic head at exit point of Buffalo Creek is 194.3351 m. This comes out to be about 0.0476 % hydraulic head gradient which is in agreement with ground surface slope. This difference is considered as acceptable; the 200 m × 200 m cell size ground surface elevation value is considered by GIS to be a surrogate elevation for the entire cell. Another output result by MODFLOW is the annual volumetric water budget balance, (input–output). This percentage difference is 0 %, indicating good result. Volumetric water budget output by MODFLOW is also compared and verified with the separated (baseflow separation model) and calculated value of baseflow. MODFLOW calculates 3.4 % by volume less than the calculated water balance.

References

Arnold, J. G., & Allen, P. M. (1999). Validation of automated methods for estimating baseflow and groundwater recharge from streamflow records. *Journal of American Water Resources Association, 35*, 411–424.

Chiang, W.-H., & Kinzelbach, W. (2001). *Groundwater Modeling with PMWIN*. New York: Springer. doi:10.1007/978-3-662-05551-6. ISBN 978-3-662-05551-6.

City of Greensboro. (2011). Sewage collection and reclamation plant. City of Greensboro Open-File- Report. Retrieved from www.greensboro-nc.gov/Modules/ShowDocument.aspx?documentID.

Daniel, C. C., & Harned, D. A. (1998). Ground-water recharge to and storage in the regolith-fractured crystalline rock aquifer system, Guilford County, North Carolina. US Geological Survey Water-Resources Investigations Report 97–4140.

Harbaugh, A. W., Banta, E. R., Hill, M. C., & McDonald, M. G. (2000). MODFLOW-2000, The U.S. geological survey modular ground-water model—User guide to modularization concepts and the ground-water flow process: Open-File Report 00-92, 121p. Retrieved from http://water.usgs.gov/ogw/MODFLOW_list_of_reports.html.

HERA Team. (2007). Guilford County groundwater monitoring. Department of Public Health Report. Open-File Report.

Kimberly, M. Y. (2002). Functional assessment for proposed flood plain storm water treatment wetland.

Kopec, R. J., & Clay, J. W. (1975). Guilford County groundwater monitoring. Department of Public Health: Status Report. Retrieved from www.co.guilford.nc.us/gheh_cms/hera/forms/NetworkReport.pdf.

Lyne, V. D., & Hollick, M. (1979). Stochastic time variable rainfall-runoff modeling. In *Hydro and Water Resource Symposium* (pp. 89–92). Perth, Australia: Institution of Engineers Australia.

McDonald, M. G., & Harbaugh, A. W. (1988). A modular three-dimensional finite-difference groundwater flow model. Techniques of water-research investments of the US Geological Survey, Book 6, Chapter A1.

Nathan, R. J., & McMahon, T. A. (1990). Evaluation of automated techniques for baseflow and recession analysis. *Water Resources Research, 26*, 1465–1473.

Prudic, D. E. (1989). Documentation of a computer program to simulate stream-aquifer relations using a modular, finite difference groundwater flow model. US Geological Survey, Open-File Report, 88–729.

Part III
Food Bioprocessing

Inactivation of *E. coli* O157:H7 on Rocket Leaves by Eucalyptus and Wild-Thyme Essential Oils

Saddam S. Awaisheh

Abstract Recently, the consumption of fresh produce has been associated with the outbreaks of *Escherichia (E.) coli* O157:H7, the dangerous foodborne pathogen. Accordingly, the objective of this work was to evaluate the effectiveness of essential oils (EOs) of 24 different plants and herbs in controlling growth of *E. coli* O157:H7 in rocket leaves model. EOs antibacterial activities and MICs were determined by disk diffusion and micro-dilution methods, respectively. The effectiveness of most active EOs (0.5 and 1 % v/w) were further examined for sanitizing rocket leaves contaminated with 3 log CFU/leaf of *E. coli* O157:H7 strains cocktail after storage at 4 °C/3 days and 25 °C/24 h. The EOs of Wild-thyme and Eucalyptus had the most ($p < 0.05$) inhibition activity with Minimal Inhibitory Concentration (MIC) values of 0.32 µL/g. In control rocket leaves samples, *E. coli* O157:H7 populations increased by 1.16 and 2.60 log cycles after storage at 4 °C/3 days and 25 °C/24 h, respectively. Compared with control samples, sanitizing of rocket leaves stored at 25 °C/24 h with either 0.5 or 1 % of Eucalyptus or Wild-thyme resulted in 1.85, 1.97, 1.95, and 1.98 log cycles reduction in population counts, respectively, whereas sanitizing of samples stored at 4 °C/3 days resulted in 1.82, 1.85, 1.90, and 1.99 log cycles reduction, respectively. Findings showed that EOs of Wild-thyme and Eucalyptus were effective in sanitizing *E. coli* O157:H7 on rocket leaves at both storage conditions. Our results suggest that these EOs could be used as natural sanitizers in many final produce products, with better safety profile and consumer acceptance compared to chemical sanitizers.

Introduction

Fresh produce (FP) refers to fresh fruits and vegetables that have been minimally processed and altered in form of peeling, slicing, chopping, shredding, coring, or trimming, with or without washing (Brackett 1999). The increasing consumer

---

S.S. Awaisheh (✉)
Department of Nutrition and Food Processing, Al Balqa Applied University,
Salt 19117, Jordan
e-mail: saddam_awaisheh@yahoo.com

demands for fresh, healthful, convenient, and ready-to-eat (RTE) products have resulted in a rapid expansion in the FP industry and market (Kim and Harrison 2008). In parallel to the expansion of FP market, foodborne outbreaks associated with FP consumption have increased dramatically (Olaimat and Holley 2012). In 2009, leafy green vegetables were ranked as the first among the top ten riskiest foods regulated by the FDA, based on the cases of illnesses reported between 1990 and 2007 (CDC 2009). The FP contamination with harmful bacteria could result from pathogens found in the soil or water where produce grows, or during postharvesting practices, such as preparation, packaging, or storage; or from contaminated wash water, containers, processing equipment; or improper handling by workers (Kim and Harrison 2008; Brackett 1999).

Several studies reported the incidence of various harmful foodborne pathogens in virtually all types of fresh produce. For example, Brackett (1999) reported the incidence of different types of pathogens, including Enterotoxigenic and Enterohemorrhagic *Escherichia coli*, *Salmonella* spp., *Listeria monocytogenes*, *Shigella* spp., *Vibrio* spp., as well as *Campylobacter* spp., in different types of raw fruits and vegetables. Among the different pathogens, *E. coli* O157:H7 has been considered as one of the most serious pathogens that occur in FP (Al-Nabulsi et al. 2013). It has been frequently associated with several foodborne outbreaks caused by consumption of leafy greens like lettuce, spinach, and celery. Of these outbreaks, 23 which occurred between 1995 and 2007 were associated with the consumption of lettuce and spinach (CDC 2009). Another major outbreak in 15 countries in Europe and North America with nearly 4000 illnesses was reported in 2011, resulted from the consumption of sprouted fenugreek seeds contaminated with *E. coli* O104:H4 (ECDC 2011).

Different disinfection procedures have been applied to reduce the incidence and consequently the risks associated with the consumption of green leafy vegetables contaminated with foodborne pathogens, including *E. coli* O157:H7. The use of chemical disinfectants is the most common procedure, such as chlorine dioxide in the gaseous or liquid forms, peroxyacetic acid, acidic electrolyzed water, and other approaches (Mahmoud and Linton 2008; Olaimat and Holley 2012). However, each chemical disinfectant type has its own limitations. For example, chlorine, the most used, is inactivated when comes in contact with organic material and can form compounds with negative effects on human health (Stevens 1982). This has highlighted the need for alternative safe disinfection compounds. Plant essential oils (EOs) are among the most recommended candidates (Awaisheh 2013; Burt 2004), and they are of increasing interest because of their antibacterial, antifungal, and antioxidant properties (Alzoreky and Nakahara 2003; Beuchat and Golden 1989). So, considerable efforts are being made to search for antimicrobial agents from medicinal plant species (Awaisheh 2013). Therefore the objective of this work was to study the effectiveness of EOs of different medicinal plants to detach or inactivate the growth of *E. coli* O157:H7 at the surface of rocket salad leaves.

Materials and Methods

Pathogen strains and inoculum preparation. Four nonpathogenic (verotoxigenic negative) clinical isolates of *E. coli* O157:H7 (0304, 0627, 0628, and 3581) were provided by Rafiq Ahmed, National Microbiology Laboratory, Public Health Agency, Canadian Science Centre for Human and Animal Health (Winnipeg, MB, Canada). All strains were prepared stored, and activated according to the procedure described by Al-Nabulsi et al. (2013). Briefly, strains were stored individually in trypticase soy broth (TSB) (Oxoid Ltd., Basingstoke, UK) containing 20 % (v/v) glycerol (Sigma-Aldrich, St. Louis, MO, USA) at −40 °C. For reactivation, a loopful from each culture was inoculated into 10 mL sterile TSB and incubated at 37 °C for 24 h. Then strains were stored at 4 °C in Sorbitol MacConkey (SMAC) agar. Prior to the experiment, the strains were individually inoculated into 10 mL sterile TSB and incubated at 37 °C for 24 h. The cocktail of the strains was prepared by mixing 2.5 mL of each culture. This cocktail was centrifuged out and pellets were resuspended with sterile deionized water to achieve culture concentration of 10^8 CFU/mL (Obaidat and Frank 2009).

Medicinal plants and essential oils extraction. The pure EOs of 22 different endemic medicinal plants (Table 1) in Jordan were procured from a local extraction facility in Amman-Jordan. The EOs were extracted in the facility by using

Table 1 List of scientific and common names of medicinal plants used in extraction of Eos

Scientific name	Common name
Mentha arvensis	Wild-mint
Carum carvi	Caraway
Salvia officinalis	Sage
Foeniculum vulgare	Fennel
Laurus nobilis	Laurel
Cinnamomum verum	Cinnamon
Rosmarinus officinalis	Rosemary
Eucalyptus deglupta	Eucalyptus
Thymus serpyllum	Wild-thyme
Zingiber officinale	Ginger
Pimpinella anisum	Anise
Trifolium	Clove
Elettaria	Cardamom
Nigella sativa	Cumin
Coriandrum sativum	Coriander
Achillea arabica	Qysoom
Artemisia absinthium	Booh
Abies cilicica	Fir
Chamaemelum nobile	Chamomile
Trigonella foenum-fraecum	Fenugreek
Capsicum annuum	Hot red chili

steam-distillation instrument. It was then kept in a sealed dark glass vial at 4 °C until used. Serial twofold dilutions were prepared by adding equal quantities of 0.5 % dimethyl sulfoxide (DMSO) to the EOs just prior to the experiment.

Screening of essential oils antibacterial activity. The antibacterial activity of the 22 EOs against the selected *E. coli* O157:H7 strains alone or combined has been determined by the disk diffusion method (Awaisheh 2013; Klančnik et al. 2010). Aliquots of 20 μL of each EO were dispensed onto 6-mm-diameter sterile filter paper discs with air drying. Each of the discs was placed on trypticase soy agar (TSA) that had been previously seeded with *E. coli* O157:H7 strains alone or combined and incubated at 37 °C for 24 h. Inhibitory zones were measured in millimeter using a caliper. The synergistic activity of mixture of the EOs that showed the maximum antibacterial activity (EOs of Wild-thyme and Eucalyptus) was done using the same protocol. The EOs mixture was made by mixing equal portions of each EO in DMSO, then 20 μL of the mixture was applied onto 6-mm-diameter sterile filter paper.

Antibacterial activity assay. The minimal inhibitory concentration (MIC) values were determined for the EOs of Wild-thyme or Eucalyptus. MIC values were determined using 96-wells micro-dilution method (Awaisheh 2013; Yasunaka et al. 2005). A single colony of each *E. coli* O157:H7 strain was separately transferred into a test TSB tube and incubated at 37 °C for 24 h. A 100 μL of media containing 4.3 log CFU of the bacterium was added to corresponding wells. The range of 0.08–100 μL/mL for each EO was prepared by twofold serial dilutions of pure EOs by using DMSO. A 100 μL of the suitable EOs dilution was added to corresponding wells. The total volume in each well was 200 μL. The sealed plates were shaken and incubated at 37 °C for 24 h. Microbial growth in each well was determined by reading the respective absorbance (Abs) at 600 nm using microplate reader (ELX 800, Biotek, Highland Park, VT, USA). The positive and negative controls used were pure DMSO and EOs containing DMSO, respectively. MIC values represented the lowest EOs dilution at which the growth is absent, and this was confirmed by plating 20 μL aliquots from the clear wells onto SMAC plates.

Preparation and inoculation of rocket salad leaves. Rocket leaves were purchased from a local market in Amman-Jordan, on the day of experiment. Preparation of rocket leaves was done according to the procedure describes by Al-Nabulsi et al. (2013) with some modifications. Briefly, damaged leaves were removed; undamaged leaves were washed with sterilized water, dried using a salad spinner, and then exposed to UV light ($\lambda = 260$ nm) for 10 min to kill surface contaminants, followed by plating part of the samples onto SMAC and general purpose nutrient agar plates (Oxoid, Basingstoke, UK) to confirm the leaves sterility. The surface of sterile rocket leaves was inoculated by adding 50 μL, containing 3 log CFU of *E. coli* O157:H7 cocktail, at 5 different locations/leaf (on the upper surface). Then inoculated leaves were dried in a biosafety cabinet for 2 h at 25 °C to allow attachment of bacterial cells to the leaf surface (Lang et al. 2004). Test leaves were then stored at 25 °C/24 h or at 4 °C/3 days.

Efficacy of EOs solutions to sanitize E. coli O157:H7 on rocket leaf surfaces. Two different concentrations, 0.5 and 1 %, of Wild-thyme and Eucalyptus EOs solutions alone and in combination were prepared. The EOs sanitizing solutions were composed of 0.9 % saline and 1 % tween 80 dissolved in sterile deionized water. The EOs sanitizing solutions were freshly prepared on the day of experiment. After storage at 25 °C/24 h or 4 °C/3 days, each inoculated rocket leaf ($\simeq$0.85 g) was placed into a stomacher bag containing 100 mL of each sanitizing solution for 5 min with agitation. The leaves were transferred to a salad spinner to remove excess liquid from leaf surfaces (Al-Nabulsi et al. 2013; Keskinen et al. 2009). The leaves were analyzed for the presence of *E. coli* O157:H7, by transferring the rocket leaf into a sterile stomacher bag containing 100 mL of 0.1 % peptone water solution and serially diluted in the same solution. From appropriate dilutions 0.1 mL was plated on the surface of SMAC agar supplemented with 0.05 mg/L Cefixim and 2.5 mg/L Potassium Tellurite and overlaid with TSA to facilitate the growth of injured cells. The plates were then incubated at 37 °C for 18–24 h (Keskinen et al. 2009; Osaili et al. 2010). Inoculated leaves stored at both storage conditions with no EOs treatments served as controls.

Statistical analysis. All data were analyzed using one-way analysis of variance of a Statistical Analysis System (SAS) software version 9.3 (SAS 2011). A Tukey–Kramer test was performed to compare any significant differences ($p < 0.05$, unless otherwise indicated) in variables between groups. The data normal distribution was analyzed using SAS Univariate Procedure. Each experiment was repeated three times.

Results

Antibacterial activity assay. In this study the antibacterial activity of EOs of 22 different plants were screened against four *E. coli* O157:H7 strains, alone and combined. The antibacterial activity of these EOs against *E. coli* O157:H7 strains were presented as the diameter of inhibition zone (Table 2). The inhibition zones with 20 μL of these EOs ranged from 0 to 40 mm in diameter. Four modes of antibacterial activity were demonstrated: (1) Absence of activity (data not shown); (2) weak activity (1–2 mm inhibition zone) (data not shown); (3) mild activity (5–10 mm inhibition zone); and (4) strong activity (10–40 mm inhibition zone). Among the 22 screened EOs, only 8 EOs were shown to have a mild activity, whereas 2 EOs were shown to have a strong activity (Table 2). The EOs with the mild antibacterial activity included Fennel, Fir, Wild-mint, Coriander, Qysoom, Caraway, Rosemary, and Cinnamon, with inhibition zones ranging from 5.5 to 18 mm in diameter. The EOs with the strongest activity were Wild-thyme and Eucalyptus. The Wild-thyme inhibition zones ranged from 36 to 40 mm against the individual strains and 38 mm against the cocktail of the strains, whereas for Eucalyptus, it ranged from 25.5 to 32 mm against the individual strains and 30 mm against the cocktail of the strains. Results showed that the antibacterial activity of Wild-thyme and Eucalyptus was significantly ($p < 0.05$) higher than the rest EOs, and that of Wild-thyme was significantly ($p < 0.05$) higher than Eucalyptus.

Table 2 Essential oils of medicinal plants with mild to strong antibacterial activity against individual and cocktail of four strains of *E. coli* O157:H7

Strains of *E. coli* O157:H7	Essential oils (20 µL)								
	Wild-thyme	Eucalyptus	Fennel	Fir	Wild-mint	Coriander	Qysoom	Caraway	Rosemary	Cinnamon
Zone of inhibitions (mm)										
0304	38 ± 3.6a[2,3]	32 ± 2.8a	10.7 ± 0.1c	12 ± 0.5b	8.7 ± 0.3d	13.5 ± 0.3b	12 ± 0.4b	10.5 ± 0.6c	12 ± 0.4b	5.5 ± 0.3e
0627	40 ± 2.8a	25.5 ± 1.3b	5.5 ± 0.7e	15 ± 0.9c	8 ± 0.5d	11 ± 1.3d	9.5 ± 0.9d	7.5 ± 0.9a	9 ± 0.9d	5.5 ± 0.3e
0628	36 ± 2.6a	30 ± 1.2b	10 ± 0.9d	10 ± 0.7d	15.1 ± 1.8c	0.0f	12 ± 0.5c	12.5 ± 1.8c	13.5 ± 1.8c	6.5 ± 0.5e
3581	39 ± 2.7a	26 ± 2.5b	22 ± 2.1c	12 ± 0.5d	18 ± 1.4c	11 ± 1.3d	10 ± 0.9a	12.5 ± 0.7d	12 ± 1.2d	8 ± 1.5e
Cocktail	38 ± 2.2a	30 ± 1.7b	16 ± 2.7c	13 ± 1.3c	13 ± 1.7c	10 ± 1.1d	10 ± 1.0d	9 ± 0.7d	9 ± 0.6a	5.5 ± 0.7e

[1]Antibacterial activity were screened using disk diffusion method and measured as a diameter of inhibition zone in millimeter (Awaisheh 2013; Klančnik et al. 2010)
[2]Results are means ± standard deviation of four determinations
[3]Means within the same row with different letters are significantly different ($p < 0.05$)

MIC values of Wild-thyme and Eucalyptus EOs. The strong antibacterial activity showed by EOs of Wild-thyme and Eucalyptus against four strains of *E. coli* O157:H7 was confirmed by the micro-dilution method assay to determine MIC values. Table 3 shows the MIC values of EOs of Wild-thyme and Eucalyptus, and these values were 0.312 for both EOs. The low MIC values confirmed that both EOs exhibited significantly ($p < 0.05$) equal bactericidal effects on the tested strains.

Effectiveness of EOs sanitizing interventions against cocktail of E. coli O157:H7 strains. The effect of EOs (Eucalyptus and Wild-thyme) on the contamination level of 3 log CFU/g of four strains cocktail of *E. coli* O157:H7 at the surface of rocket leaves model during storage at 25 °C/24 h and 4 °C/3 days is shown in Table 4.

However, data showed that the inoculated cells successfully attached to the surface of rocket leaves. Table 4 showed that a 2.5 and 1.06 log CFU/leaf increase in the numbers of inoculated cells was noted when rocket leaves were stored at

Table 3 MIC[1] values of Wild-thyme or Eucalyptus EOs against four strains of *E. coli* O157:H7

Strains of *E. coli* O157:H7	MIC values of EOs (μL/mL)[2]	
	Wild-thyme	Eucalyptus
EC 0304	0.312 ± 0.0a[3]	0.312 ± 0.0a
EC 0627	0.312 ± 0.0a	0.312 ± 0.0a
EC 0628	0.312 ± 0.0a	0.312 ± 0.0a
EC 3581	0.312 ± 0.0a	0.312 ± 0.0a

[1]MIC values were screened using wells micro-dilution method
[2]Readings are means of four determinations ± standard deviation
[3]Results with different letters in the same row of the MIC values are significantly different ($p < 0.05$)

Table 4 Recovery of cells of cocktail of four strains of *E. coli* O157:H7 on the surface of rocket leaves stored at 25 °C/24 h and 4 °C/3 days after sanitizing with two concentrations of Wild-thyme or Eucalyptus EOs solutions

EOs treatment	Initial counts[1]	Storage conditions	
		25 °C/24 h	4 °C/3 days
Control	3.1 ± 0.02a[2] (c)[3]	5.60 ± 0.29a (a)	4.16 ± 0.29a (b)
0.5 % Wild-thyme	3.2 ± 0.19a (b)	4.61 ± 0.21b (a)	3.48 ± 0.21b (c)
		(1.95 Log)[4]	(1.90 Log)
1.0 % Wild-thyme	3.12 ± 0.22a (b)	4.11 ± 0.17c (a)	2.30 ± 0.12b (c)
		(1.99 Log)	(1.99 Log)
0.5 % Eucalyptus	3.09 ± 0.23a (c)	5.06 ± 0.36a (a)	3.71 ± 0.22a (b)
		(1.85 Log)	(1.82 Log)
1.0 % Eucalyptus	3.11 ± 0.01a (b)	4.46 ± 0.12b (a)	3.63 ± 0.15c (c)
		(1.97 Log)	(1.85 Log)

[1]Microbial count is expressed as log CFU/leaf, readings are means of four determinations ± standard deviation
[2]Results with different letters in the same column are significantly different ($p < 0.05$)
[3]Results with different letters in brackets in the same row are significantly different ($p < 0.05$)
[4]Level of inhibition: expressed as log cycle reduction in bacterial count in the treated samples to the count in the respective control

25 °C/24 h and 4 °C/3 days, respectively. Variable decontamination levels were obtained for rocket leaves by two concentrations of Wild-thyme and Eucalyptus EOs. The decontamination effectiveness ranged from reduction of 1.82 to 1.99 log cycles. Decontamination effectiveness of EOs of Wild-thyme were 1.95 and 1.98; and 1.90 and 1.99 log cycles reduction for the 0.5 and 1 % EO concentrations at 25 °C/24 h and 4 °C/3 days, respectively. Whereas decontamination level of Eucalyptus EOs were 1.85 and 1.97; and 1.82 and 1.85 log cycle reduction for the 0.5 and 1 % EO concentrations at 25 °C/24 h and 4 °C/3 days, respectively. Results demonstrated that EO of Wild-thyme was significantly higher ($p < 0.05$) and more effective in decontaminating rocket leaves than EO of Eucalyptus at both storage conditions. Similarly, results demonstrated that the 1 % concentration was more effective for both EOs than the 0.5 % concentration at both storage conditions.

Discussion

The change in consumer habits towards increased consumption of leafy vegetables, as part of healthy diet, has been accompanied by increased numbers of foodborne outbreaks associated with fruits and vegetables. Rocket leaves are popular leafy greens consumed as an ingredient in salads or consumed alone as a side dish in Jordan and many other countries around the world. Various studies reported the incidence of foodborne pathogens in rocket leaves. For example, Russell et al. (2010) reported that the contamination of rocket leaves was 100 % with *E. coli* at a level of ~6 log CFU/g. Nygard et al. (2008) reported that rocket leaves contaminated with *Salmonella enterica* were involved in an outbreak that took place in the UK and Scandinavia in 2004. However, several chemical disinfectants are used to reduce the numbers of attached viable *E. coli* O157:H7 on the surface of FP, including chlorine solution, peroxyacetic acid, ozone, chlorine oxide, acidified sodium chlorite, sodium chlorite, sodium hypochlorite, and citric acid (Allende et al. 2009; Al-Nabulsi et al. 2013; Singh et al. 2002). Several limitations and health consequences have been reported for these disinfectants. Accordingly, plant EOs have been proposed as a suitable and safe alternative for these disinfectant. In vitro, many EOs have been studied and approved for their antibacterial activity against *E. coli* O157:H7 including those of Thyme, Cinnamon, Marjoram, Clove, Garlic, Oregano, Rosemary, etc. (Holley and Patel 2005; Alzoreky and Nakahara 2003; Burt and Reinders 2003; Friedman et al. 2002; Singh et al. 2002). Even though Thyme EO antibacterial activity have been extensively studied and approved against wide range of pathogens, literature regarding the antibacterial activity of Eucalyptus is very limited. However, our results revealed a strong inhibitory effect of EOs of Wild-thyme and Eucalyptus against *E. coli* O157:H7. In consistent to our findings, Burt and Reinders (2003) reported that Oregano and Thyme EOs had the strongest bacteriostatic and bactericidal effects against *E. coli* O157:H7, followed by Bay and Clove buds. Furthermore, Friedman et al. (2002) reported that among 96 EOs, the EOs of Oregano, Thyme, Cinnamon, Palmarosa, Bay leaf, Clove bud, and Lemon grass had the maximum inhibitory effect against *E. coli*. Yossa et al. (2010) reported the control of *E. coli* growth by the EOs of Cinnamaldehyde and Eugenol. Also, Selim (2011) reported the effectiveness of Thyme EO, and the low effectiveness of Eucalyptus EO in control of *E. coli* O157:H7 in feta cheese and minced beef.

Current results revealed a rapid growth rate of *E. coli* O157:H7 in the rocket leaves upon storage, with higher growth rate at 25 °C/24 h than 4 °C/3 days. These results could indicate the possibility of the increase of *E. coli* O157:H7 numbers in rocket leaves under the normal storage conditions encountered in the market and at home levels. Also, these results indicated the necessity for the sanitizing treatment of various FP products before consumption. However, our results showed that washing of rocket leaves with EOs solutions, particularly 1 % of Wild-thyme and Eucalyptus, has resulted in 1.98 and 1.99; and 1.97 and 1.85 log cycle reduction in *E. coli* O157:H7 count after storage at 25 °C/24 h than 4 °C/3 days, respectively. Lower reduction levels were revealed at 0.5 % EOs concentration. Upon search in literature, very limited reports were found regarding the study of the efficacy of EOs as sanitizing solution in FP (Singh et al. 2002; Wan et al. 1998). Comparable to our findings, Singh et al. (2002) reported a significant reduction (1.65 and 1.90 log CFU/g, respectively) in the count of *E. coli* O157:H7 on lettuce and baby carrots when using thyme oil. Earlier, Wan et al. (1998) reported that washing lettuce with 1.0 and 10.0 mL/L suspensions of Basil EO had resulted in 2.0 and 2.3 log cycle reductions of viable bacteria on lettuce, respectively. The promising antibacterial activity of Wild-thyme and Eucalyptus EOs could be attributed to their high contents of thymol and 1,8-cineol, respectively. These compounds have been known for their antibacterial activities (Dorman and Deans 2000; Burt and Reinders 2003; Selim 2011).

The washing of FP with chemical disinfectant is the common sanitizing practice. However, compared to current findings, in our previous work (Al-Nabulsi et al. 2013), a ≤ 1.0 log reduction (CFU/leaf) in count of *E. coli* O157:H7 upon washing rocket leaves with 200 ppm chlorine was reported. Also, Aruscavage et al. (2006) reported that free chlorine at concentrations of 20–200 ppm did not eradicate pathogens from the surface of fresh produce, but a reduction of 1–3 log CFU/g was attained. Additionally, Park and Beuchat (1999) reported that the population of *E. coli* O157:H7 and *Salmonella* on intact melon surfaces was reduced by 2.6–3.8 log CFU/g when washed with peroxyacetic acid at 40–80 ppm.

Although the *E. coli* O157:H7 cocktail was neither eliminated nor completely inhibited, EOs treatments were able to reduce population of *E. coli* O157:H7 cocktail at both storage conditions, in a magnitude that is comparable to the most common disinfectants used in FP sanitizing, namely chlorine and peroxyacetic acid. However, the failure of most of the chemical disinfectants and EOs treatment in totally eliminating *E. coli* O157:H7 cells from the surfaces of FP is referred to the attachment mechanism by which *E. coli* O157:H7 cells attach to the surfaces of FP. This might be explained in part by the irreversible attachment of *E. coli* O157:H7 to the leaf surface. In addition, cells of *E. coli* O157:H7 may locate themselves in inaccessible sites like crevices and pores, punctures, and other damaged tissue areas (Seo and Frank 1999; Burnett et al. 2000). Another possible explanation is that cells were able to form biofilms that act as barriers which limit the effectiveness of sanitizers (Solomon et al. 2005). Ryu and Beuchat (2005) indicated that *E. coli* O157:H7 forms biofilms on the surface of stainless steel which leads to increased resistance to chlorine.

Conclusion

The current findings indicated that EOs solutions comparable to routine chemical disinfectant reduced *E. coli* O157:H7 on the surfaces of rocket leaves. Also, it suggests that the EOs of Wild-thyme and Eucalyptus has good antibacterial potential in food preservation with better acceptance to consumers and the regulatory agencies than chemical compounds. Also, the synergistic activity and compatibility of Wild-thyme and Eucalyptus EOs could be considered as potential candidates for further studies.

References

Allende, A., McEvoy, J., Tao, Y., & Luo, Y. (2009). Antimicrobial effect of acidified sodium chlorite, sodium chlorite, sodium hypochlorite, and citric acid on Escherichia coli O157:H7 and natural microflora of fresh-cut cilantro. *Food Control, 20*, 230–234.

Al-Nabulsi, A. A., Osaili, T. M., Obaidat, H. M., Shaker, R. R., Awaisheh, S. S., & Holley, R. A. (2013). Inactivation of stressed E. coli O157:H7 cells on the surface of rocket salad leaves by chlorine and peroxyacetic acid. *Journal of Food Protection, 77*(1), 32–39.

Alzoreky, S., & Nakahara, K. (2003). Antimicrobial activity of extracts from some edible plants commonly consumed in Asia. *International Journal of Food Microbiology, 80*, 223–230.

Aruscavage, D., Lee, K., Miller, S., & LeJeune, J. T. (2006). Interactions affecting the proliferation and control of human pathogens on edible plants. *Journal of Food Science, 71*, R89–R99.

Awaisheh, S. (2013). Efficacy of Fir and Qysoom essential oils, alone and in combination, in controlling *Listeria monocytogenes* in vitro and in RTE meat products model. *Food Control, 34*, 657–661.

Beuchat, R., & Golden, A. (1989). Antimicrobials occurring naturally in foods. *Food Technology (Chicago), 43*, 134–142.

Brackett, R. E. (1999). Incidence, contributing factors, and control of bacterial pathogens in produce. *Postharvest Biology and Technology, 15*, 305–311.

Burnett, S. L., Chen, J., & Beuchat, L. R. (2000). Attachment of *Escherichia coli* O157:H7 to the surfaces and internal structures of apples as detected by confocal scanning laser microscopy. *Applied and Environmental Microbiology, 66*, 4679–4687.

Burt, S. (2004). Essential oils: Their antibacterial properties and potential applications in foods-a review. *International Journal of Food and Microbiology, 94*, 223–253.

Burt, S. A., & Reinders, R. D. (2003). Antibacterial activity of selected plant essential oils against Escherichia coli O157:H7. *Letters in Applied Microbiology, 36*, 162–167.

Centers for Disease Control and Prevention (CDC). (2009). Surveillance for foodborne disease outbreaks—United States. Retrieved March 23, 2013, from http://www.cdc.gov/mmwr/preview/mmwrhtml/mm6035a3.htm?s_cid=mm6035a3_w

Dorman, H. J. D., & Deans, S. G. (2000). Antimicrobial agents from plants: Antibacterial activity of plant volatile oils. *Journal of Applied Microbiology, 88*, 308–316.

European Centre for Disease Prevention and Control (ECDC). (2011). Shiga toxin-producing *E. coli* (STEC): Update on outbreak in the EU (27 July 2011, 11:00). Retrieved June 11, 2013, from http://www.ecdc.europa.eu/en/activities/sciadvice/Lists/-bf0f23083f30&ID¼1166& RootFolder¼%2Fen%2Factivities%2Fsciadvice%2FLists%2FECDC%20Reviews

Friedman, M., Henika, P. R., & Mandrell, R. E. (2002). Bactericidal activities of plant essential oils and some of their isolated constituents against *Campylobacter jejuni, Escherichia coli, Listeria monocytogenes*, and *Salmonella enterica. Journal of Food Protection, 65*(10), 1545–1560.

Holley, A., & Patel, D. (2005). Improvement in shelf-life and safety of perishable foods by plant essential oils and smoke antimicrobials. *Journal of Food Microbiology, 22*, 273–292.

Keskinen, L. A., Burke, A., & Annous, B. A. (2009). Efficacy of chlorine, acidic electrolyzed water and aqueous chlorine dioxide solutions to decontaminate *Escherichia coli* O157:H7 from lettuce leaves. *International Journal of Food Microbiology, 132*, 134–140.

Kim, J. K., & Harrison, M. A. (2008). Transfer of *Escherichia coli* O157:H7 to romaine lettuce due to contact water from melting ice. *Journal of Food Protection, 71*, 252–256.

Klančnik, A., Piskernik, S., Jeršek, B., & Mozina, S. S. (2010). Evaluation of diffusion and dilution methods to determine the antibacterial activity of plant extracts. *Journal of Microbiological Methods, 81*, 121–126.

Lang, M. M., Harris, L. J., & Beuchat, L. R. (2004). Evaluation of inoculation method and inoculum drying time for their effects on survival and efficiency of recovery of *Escherichia coli* O157:H7, *Salmonella*, and *Listeria monocytogenes* inoculated on the surface of tomatoes. *Journal of Food Protection, 67*, 732–741.

Mahmoud, B. S. M., & Linton, R. H. (2008). Inactivation kinetics of inoculated *Escherichia coli* O157:H7 and *Salmonella enterica* on lettuce by chlorine dioxide gas. *Food Microbiology, 25*, 244–252.

Nygard, K., Lassen, J., Vold, L., Andersson, Y., Fisher, I., Löfdahl, S., et al. (2008). Outbreak of *Salmonella* Thompson infections linked to imported rucola lettuce. *Foodborne Pathogens and Disease, 5*, 165–173.

Obaidat, M. M., & Frank, J. F. (2009). Inactivation of *Escherichia coli* O157:H7 on the intact and damaged portions of lettuce and spinach leaves using allyl isothiocyanate, carvacrol, and cinnamaldehyde in vapor phase. *Journal of Food Protection, 72*, 2046–2055.

Olaimat, A. N., & Holley, R. A. (2012). Factors influencing the microbial safety of fresh produce: A review. *Food Microbiology, 32*, 1–19.

Osaili, T. M., Al-Nabulsi, A. A., Shaker, R. R., Al-Holy, M. M., Al-Haddaq, M. S., Olaimat, A. N., et al. (2010). Efficacy of the thin agar layer method for the recovery of stressed *Cronobacter* spp. (*Enterobacter sakazakii*). *Journal of Food Protection, 73*, 1913–1918.

Park, C. M., & Beuchat, L. R. (1999). Evaluation of sanitizers for killing *Escherichia coli* O157: H7, *Salmonella* and naturally occurring microorganisms on cantaloupes, honeydew melons, and asparagus. *Dairy, Food and Environmental Sanitation, 19*, 842–847.

Russell, D. J., Abdul Majid, S., & Tobias, D. (2010). The presence of persistent coliform and *E. coli* contamination sequestered within the leaves of the popular fresh salad vegetable "Jarjeer/Rocket" (*Eruca sativa* L.). *Egyptian Academic Journal of Biological Sciences, 2*, 1–8.

Ryu, J. H., & Beuchat, L. R. (2005). Biofilm formation by *Escherichia coli* O157:H7 on stainless steel: Effect of exopolysaccharide and Curli production on its resistance to chlorine. *Applied and Environmental Microbiology, 71*, 247–254.

Selim, S. (2011). Antimicrobial activity of essential oils against vancomycin-resistant *Enterococci* (VRE) and *E. coli* O157:H7 in feta soft cheese and minced beef. *Brazilian Journal of Microbiology, 42*, 187–196.

Seo, K. H., & Frank, J. F. (1999). Attachment of *Escherichia coli* O157:H7 to lettuce leaf surface and bacterial viability in response to chlorine. *Journal of Food Protection, 62*, 3–9.

Singh, N., Singh, R. K., Bhunia, A. K., & Stroshine, R. L. (2002). Efficacy of chlorine dioxide, ozone, and thyme essential oil or a sequential washing in killing Escherichia coli O157:H7 on lettuce and baby carrots. *Food Science and Technology, 35*, 720–729.

Solomon, E. B., Niemira, B. A., Sapers, G. M., & Annous, B. A. (2005). Biofilm formation, cellulose production, and Curli biosynthesis by *Salmonella* originating from produce, animal, and clinical sources. *Journal of Food Protection, 68*, 906–912.

Statistical Analysis System (SAS). (2011). *SAS software version 9.3*. Cary, NC: SAS Institute.

Stevens, A. (1982). Reaction products of chlorine dioxide. *Environmental Health Perspectives, 46*, 101–110.

Wan, J., Wilcock, A., & Coventry, M. J. (1998). The effect of essential oils of basil on the growth of Aeromonas hydrophila and Pseudomonas fluorescens. *Journal of Applied Microbiology, 84*, 153–158.

Yasunaka, K., Abe, F., Nagayama, A., Okabe, H., Lozada-Pérez, L., López-Villafranco, E., et al. (2005). Antibacterial activity of crude extracts from Mexican medicinal plants and purified coumarins and xanthones. *Journal of Ethnopharmacology, 97*, 293–299.

Yossa, N., Patel, J., Miller, P., Ravishankar, S., & Lo, Y. M. (2010). Antimicrobial activity of essential oils against Escherichia coli O157:H7 inorganic soil. *Food Control, 21*, 1458–1465.

Decontamination of *Escherichia coli* O157:H7 from Leafy Green Vegetables Using Ascorbic Acid and Copper Alone or in Combination with Organic Acids

Rabin Gyawali and Salam A. Ibrahim

Abstract The objective of this study was to determine the efficacy of ascorbic acid and copper alone or in combination with organic acids in decontaminating leafy greens (cilantro, parsley, and dill) that were artificially contaminated with *Escherichia coli* O157:H7. Samples were individually submerged in an approximately 8 log CFU/mL cocktail suspension consisting of three *E. coli* O157:H7 strains. To allow attachment, inoculated samples were air dried under a biosafety hood for 2 h before exposure to various treatment solutions. Individual samples were then treated with either alone or with a combination solution of ascorbic, acetic, lactic, copper, or water (control). Our results indicated that a reduction of at least 2.0 CFU/g of *E. coli* O157:H7 was achieved when a combination of 0.2 % ascorbic with acetic acid and 50 ppm copper with 0.2 % lactic acid were applied. These results demonstrated that combination treatments could be useful in improving the microbial safety of cilantro, parsley, and dill.

Introduction

The occurrence of foodborne illnesses caused by *E. coli* O157:H7 and *Salmonella* spp. has been well documented in recent years in different fresh produce (Park et al. 2011; Kase et al. 2012; Orue et al. 2013). Washing fresh produce with running tap water may remove soil and other debris, but it has a limited effect on surface microorganisms. A variety of disinfectants have been used to reduce the bacterial populations on fruits and vegetables. However, besides their potential toxicity, these disinfectants cannot completely remove or inactivate microorganisms on

R. Gyawali
Department of Energy and Environmental Systems, North Carolina Agricultural and Technical State University, Greensboro, NC, USA

S.A. Ibrahim (✉)
North Carolina Agricultural and Technical State University, Greensboro, NC, USA
e-mail: ibrah001@ncat.edu

fresh produce (Deza et al. 2003). As a result, alternative treatments are needed to eliminate foodborne pathogens. Consumers as well as the food industry are also looking for use of natural ingredients that can help to ensure the safety of food products.

Use of several natural antimicrobials to control foodborne pathogens have been previously reported (Hayek and Ibrahim 2012; Gyawali and Ibrahim 2012, 2014; Hayek et al. 2013). Organic acid is generally recognized to be a safe chemical for use in foods and has been widely used to control the growth of pathogenic bacteria in foods (Tajkarimi et al. 2010). Lactic acid is successfully used as sanitizers on animal carcasses and may have the potential to reduce microorganisms on produce surfaces. The antimicrobial activity of organic acids and other natural compounds has already been demonstrated (Ibrahim et al. 2008; Tajkarimi et al. 2010; Hayek et al. 2013). Similarly, copper ion in low concentration is an essential micronutrient and a vital cofactor for the processing of certain enzymes. However, higher concentrations of copper can cause the inhibition, or even death, of microorganisms. Thus, there is a great potential for copper to be used as an antimicrobial agent to inhibit foodborne pathogens.

The antimicrobial activity of copper alone or in combination with lactic acid on laboratory medium and carrot juice has been reported (Ibrahim et al. 2008). Our previous study also showed that copper in combination with lactic acid significantly reduced the *E. coli* O157:H7 population on the surface of lettuce and tomatoes (Gyawali et al. 2011). This result indicates the potential application of copper and acid to decontaminate the surface of other leafy greens as well. Based on our earlier work, in this study we selected leafy greens that are usually consumed raw and that have been implicated in outbreaks of *E. coli* O157:H7. Thus, the objective of this study was to investigate the antimicrobial effect of ascorbic acid and copper alone or in combination with organic acids against *Escherichia coli* O157:H7 on leafy greens.

Materials and Methods

Bacterial strains and culture preparation. Three individual strains of *E. coli* O157:H7 (H1730, E0019, F4546) grown in BHI broth were mixed together to produce a mixture of *E. coli* O157:H7. The cells of this mixture were harvested by centrifugation at 8000 rpm for 10 min at 4 °C. The supernatant was decanted, and the cell pellet was resuspended in 500 mL of sterile peptone water (0.1 %, w/v) to give a cell number of at least 8 log CFU/mL.

Sample inoculation. Leafy greens (cilantro, dill, and parsley) were purchased from a local grocery store (Greensboro, NC) on the day of the experiments. Samples were washed with tap water, dried, and individually dipped into a bacterial solution for the inoculation for an hour. After inoculation, samples were dried under the biological safety cabinet for 2 h to facilitate the attachment of bacteria.

Treatment solution preparation. Batches of 90 mL deionized distilled water (DDW) were mixed with copper ($CuSO_4 \cdot 5H_2O$), ascorbic acid (ASC), lactic acid (LA), and acetic acid (AA) to obtain 50 ppm, 0.2 %, 0.2 %, and 0.2 %, respectively. Combinations of ASC with AA and Cu with LA were also prepared at same concentrations. An additional 90 mL of DDW without treatment has served as a control. Samples were filter sterilized using 0.2 μm Nalgene filter (NalgeNunc International, Rochester, NY, USA) before tested for antimicrobial effect. All chemicals were obtained from Thermo Fisher Scientific (Fair Lawn, NJ, USA) unless otherwise stated.

Treatment procedure. After air drying, 10 g of each sample were individually immersed in sterile plastic bags containing each treatment solution. Samples were kept in each treatment solution (control, 50 ppm Cu, 0.2 % ASC, LA, AA, or combination of Cu/LA, and Cu/AA solution) for approximately 3 min. After treating the samples, each sample was then transferred into new sterile stomacher bags containing 90 mL of peptone water and then homogenized for 1 min at 200 rpm.

Bacterial enumeration. After homogenization, solution from each treated bag was tenfold serially diluted in peptone water, and 0.1 mL of appropriate diluents was spread-plated onto BHI agar medium. The plates were incubated at 37 °C for 24 h. Colonies were counted and calculated as log CFU/g.

Statistical analysis. All experiments were conducted twice. Comparisons between treatments were analyzed by SAS 9.2 (SAS Inst., Cary, NC). *p*-Values less than 0.05 were considered statistically significant.

Results and Discussion

In this study, we investigated the effects of organic acids and Cu solution in decontaminating leafy greens artificially inoculated with *E. coli* O157:H7. Table 1 shows the effect of different treatment solutions on the *E. coli* O15:H7 population

Table 1 Effect of different treatment solution on inactivation of *E. coli* O157:H7 population (log CFU/g) on leafy greens

Treatments	Leafy greens			Average population
	Cilantro	Dill	Parsley	
Control	7.12 ± 0.04	6.91 ± 0.24	6.99 ± 0.26	7.01^A
Ascorbic acid (ASC)	6.80 ± 0.01	6.73 ± 0.04	6.45 ± 0.00	6.66^B
Acetic acid (AA)	6.42 ± 0.11	6.07 ± 0.09	6.42 ± 0.11	6.30^C
ASC + AA	5.20 ± 0.12	4.86 ± 0.02	5.39 ± 0.00	5.15^E
Lactic acid (LA)	6.08 ± 0.06	5.22 ± 0.26	6.07 ± 0.07	5.79^D
Copper (Cu)	6.56 ± 0.01	6.03 ± 0.06	6.02 ± 0.09	6.20^C
Cu + LA	5.46 ± 0.32	4.95 ± 0.02	4.98 ± 0.04	5.13^E

Values are means ± standard deviation. Values with different letters within a column are significantly different ($p < 0.05$)
Initial population was 7.16, 7.12, and 7.23 log CFU/g for cilantro, dill, and parsley, respectively (Average initial population = 7.17 log CFU/g)

on cilantro, dill, and parsley. The average initial level of *E. coli* O157:H7 was 7.17 log CFU/g (7.16, 7.12, and 7.23 log CFU/g for cilantro, dill, and parsley, respectively). Washing the sample (cilantro, dill, and parsley) with the control treatment (distilled water) reduced the cell population to only 7.01 log CFU/g on average, which indicates that water treatment is not effective in decontaminating leafy greens. There was a reduction of 0.51 and 0.87 log CFU/g when samples were treated with 0.2 % ASC or AA, respectively. However, a significant reduction ($p < 0.05$) occurred after samples were treated with a combination of ASC with AA.

On average, this combination treatment reduced the level of *E. coli* O157:H7 by 2.02 log CFU/g. A higher reduction (2.26 log) was achieved on dill leaves compared to cilantro and parsley. The antimicrobial effects of 0.2 % ASC in combination with 0.2 % LA against *E. coli* O157:H7 in culture and carrot juice medium was previously reported (Tajkarimi and Ibrahim 2011). The authors reported bacterial populations were reduced by 2.77 log CFU/mL in culture medium after 8 h of incubation at 37 °C. In carrot juice, the population was below detectable limits after 24 h of incubation. Levels of *E. coli* O157:H7 were reduced by 1.38 log CFU/g on average with 0.2 % LA. When 50 ppm Cu was used as a treatment solution, the average cell reduction was 0.97 log CFU/g. Higher cell reduction was achieved when samples were treated with Cu in combination with LA ($p < 0.05$). The *E. coli* O157:H7 population was reduced by 1.7, 2.17, and 2.25 log CFU/g on cilantro, dill, and parsley, respectively. On average, this combination treatment reduced the levels of *E. coli* O157:H7 by 2.04 log CFU/g.

Washing with water, chlorine, peroxyacetic acid, acidified sodium chlorite, hydrogen peroxide, ozone, or brush and spray washers has not always been effective enough to reduce the bacterial population in produce (Ganesh et al. 2010). For instance, the most commonly used washing treatment in the produce industry is chlorine. Foley et al. (2004) reported that the use of chlorine at 200 ppm reduced pathogenic microflora by approximately 1.5–2 logs on lettuce, cilantro, and parsley. The efficacy of the combination treatment that we used in this study was found to be superior compared to chlorine treatment. Moreover, these commonly used sanitizing solutions in the fresh produce industry have raised numerous safety concerns related to humans and the environment. In this study our treatment exposure time was approximately 3 min. It is expected that higher cells reductions could be achieved by increasing the treatment time (>3 min).

In conclusion, our results showed that organic acids and Cu have an antimicrobial effect against *E. coli* O157:H7. An average pathogens reduction of 2.03 log CFU/g was achieved on the surface of leafy greens that were treated with ASC/AA and Cu/LA combinations. These combination treatments could be useful for improving the microbial safety of cilantro, dill, and parsley and could find application in the produce industry as pre-rinsing sanitizing solutions.

Acknowledgments This publication was made possible by grant number NC.X-267-5-12-170-1 from the National Institute of Food and Agriculture and its contents are solely the responsibility of the authors and do not necessarily represent the official view of the National Institute of Food and

Agriculture. Part of this work was published in the "International Scientific Theoretical and Practical Conference of Young Scientists" (Youth and Science of the twenty-first century). The authors would like to thank Dr. Keith Schimmel for his support and Ms. Amira Ayad for her experimental assistance while conducting this work.

References

Deza, M., Araujo, M., & Garrido, M. (2003). Inactivation of *Escherichia coli* O157: H7, *Salmonella enteritidis* and *Listeria monocytogenes* on the surface of tomatoes by neutral electrolyzed water. *Letters in Applied Microbiology, 37*(6), 482–487.

Foley, D., Euper, M., Caporaso, F., & Prakash, A. (2004). Irradiation and chlorination effectively reduces *Escherichia coli* O157: H7 inoculated on cilantro (*Coriandrum sativum*) without negatively affecting quality. *Journal of Food Protection, 67*(10), 2092–2098.

Ganesh, V., Hettiarachchy, N. S., Ravichandran, M., Johnson, M. G., Griffis, C. L., Martin, E. M., et al. (2010). Electrostatic sprays of food-grade acids and plant extracts are more effective than conventional sprays in decontaminating *Salmonella* Typhimurium on spinach. *Journal of Food Science, 75*(9), M574–M579.

Gyawali, R., & Ibrahim, S. A. (2012). Impact of plant derivatives on the growth of food borne pathogens and the functionality of probiotics. *Applied Microbiology and Biotechnology, 95*(1), 29–45.

Gyawali, R., & Ibrahim, S. A. (2014). Natural products as antimicrobial agents. *Food Control, 46*, 412–429.

Gyawali, R., Ibrahim, S. A., Abu Hasfa, S. H., Smqadri, S. Q., & Haik, Y. (2011). Antimicrobial activity of copper alone and in combination with lactic acid against *Escherichia coli* O157: H7 in laboratory medium and on the surface of lettuce and tomatoes. *Journal of Pathogens*. doi:10.4061/2011/650968.

Hayek, S. A., Gyawali, R., & Ibrahim, S. A. (2013). Antimicrobial natural products. In A. Mendez-Vilas (Ed.), *Microbial pathogens and strategies for combating them: Science, technology and education*. Badajoz, Spain: Formatex Research Center.

Hayek, S. A., & Ibrahim, S. A. (2012). Antimicrobial activity of Xoconostle against *Escherichia coli* O157:H7. *International Journal of Microbiology, 2012*, 368472. doi:10.1155/2012/368472.

Ibrahim, S. A., Yang, H., & Seo, C. W. (2008). Antimicrobial activity of lactic acid and copper on growth of *Salmonella* and *Escherichia coli* O157: H7 in laboratory medium and carrot juice. *Food Chemistry, 109*(1), 137–143.

Kase, J. A., Maounounen-Laasri, A., Son, I., Deer, D. M., Borenstein, S., Prezioso, S., et al. (2012). Comparison of different sample preparation procedures for the detection and isolation of *E. coli* O157: H7 and non-O157 STECs from leafy greens and cilantro. *Food Microbiology, 32*(2), 423–426.

Orue, N., García, S., Feng, P., & Heredia, N. (2013). Decontamination of *Salmonella, Shigella,* and *Escherichia coli* O157: H7 from leafy green vegetables using edible plant extracts. *Journal of Food Science, 72*(2), M290–M296.

Park, S. H., Choi, M. R., Park, J. W., Park, K. H., Chung, M. S., Ryu, S., et al. (2011). Use of organic acids to inactivate *Escherichia coli* O157: H7, *Salmonella typhimurium*, and *Listeria monocytogenes* on organic fresh apples and lettuce. *Journal of Food Science, 76*(6), M293–M298.

Tajkarimi, M., & Ibrahim, S. A. (2011). Antimicrobial activity of ascorbic acid alone or in combination with lactic acid on *Escherichia coli* O157: H7 in laboratory medium and carrot juice. *Food Control, 22*(6), 801–804.

Tajkarimi, M., Ibrahim, S., & Cliver, D. (2010). Antimicrobial herb and spice compounds in food. *Food Control, 21*(9), 1199–1218.

Enzymatic Activity of *Lactobacillus* Grown in a Sweet Potato Base Medium

Saeed A. Hayek and Salam A. Ibrahim

Abstract The objective of this work was to study the enzymatic activity of *Lactobacillus* in a sweet potato-based medium (SPM). SPM was formed using an extract from baked sweet potatoes and supplemented with 4 g/L of each nitrogen source (beef extract, yeast extract, and proteose peptone #3). *Lactobacillus* strains were grown in SPM and MRS for 16 h at 37 °C and then plated to determine the final bacterial populations. The strains were screened spectrophotometrically for α-glucosidase, β-glucosidase, acid phosphatase, and phytase using the corresponding substrate. Our results showed no significant ($p > 0.05$) differences in the final bacterial populations of *Lactobacillus* strains grown in SPM and MRS. All *Lactobacillus* strains, except *L. reuteri*, showed similar α-glucosidase and β-glucosidase activity in SPM and MRS. *L. reuteri* showed lower α-glucosidase and higher β-glucosidase activity in SPM compared to MRS. Acid phosphatase activity of *Lactobacillus* in SPM was similar to that in MRS except for *L. reuteri* SD2112 having higher acid phosphatase in SPM than MRS. In regard to phytase, all strains showed higher activity in SPM than MRS. Strains of *L. reuteri* showed the highest enzymatic activity of α-glucosidase, acid phosphatase, and phytase whereas *L. delbrueckii* subsp. *bulgaricus* SR35 showed the highest β-glucosidase activity. Thus, the growth of *Lactobacillus* in a SPM could result in enhanced or comparable level of enzymatic activity while showing similar growth compared to MRS.

Introduction

Lactobacillus has been part of the human diet since ancient times as means of preservation especially in fermented foods. Nowadays, *Lactobacillus* is the most widely used bacteria in the food industry for fermentation and bioconversion

S.A. Hayek
Department of Energy and Environmental Systems, North Carolina Agricultural and Technical State University, Greensboro, NC, USA

S.A. Ibrahim (✉)
North Carolina Agricultural and Technical State University, Greensboro, NC, USA
e-mail: ibrah001@ncat.edu

applications. *Lactobacillus* has also been reported to be the most used probiotic bacteria due to its inherent health-promoting functionality in both humans and animals (Pfeiler and Klaenhammer 2007; Song et al. 2012). However, the use of *Lactobacillus* depends mainly on the enzymatic and metabolic activity in addition to the growth, survival, and viability of this bacteria (Pfeiler and Klaenhammer 2007; Song et al. 2012; Hayek and Ibrahim 2013). In addition, enzymes produced by *Lactobacillus* strains can alleviate some digestive problems in humans. *Lactobacillus* strains that can possess α-glucosidase, β-glucosidase, acid phosphatase, and phytase have different industrial applications and could improve nutrient availability for human health (Mahajan et al. 2010; Zotta et al. 2007; Palacios et al. 2005). α-Glucosidase (α-D-glucoside glucohydrolase, EC 3.2.1.20) is responsible for hydrolyzing glycosidic bonds in oligosaccharides (starch, disaccharides, and glycogen) and releasing α-glucose (Krasikov et al. 2001). β-Glucosidase (β-D-glucoside glucohydrolase, EC 3.2.1.21) can hydrolyse all four of β-linked glucose dimmers in cellulose to produce glucose monomers (Sestelo et al. 2004). Acid phosphatase (orthophosphoric monoester phosphohydrolase, EC. 3.1.3.2) and phytase (*myo*-inositol hexakisphosphate 6-phosphohydrolases; EC 3.1.3.26) can break down phytate and reduce its antinutritional properties (Palacios et al. 2005; López-González et al. 2008). However, these enzymes are found low in human and this may result in different health problems. Bacterial sources such as *Lactobacillus* are considered the most promising solution for such a problem.

Sweet potatoes (*Ipomoea batatas*) are an abundant agricultural product that plays a major role in the food industry and in human nutrition. There are always possible ways to enhance the contributions of sweet potatoes to the agriculture and food industries. Sweet potatoes are a rich source of carbohydrates, some amino acids, vitamins, minerals, and dietary fiber (Padmaja 2009). Sweet potatoes also contain other minor nutrients such as antioxidants, triglycerides, linoleic acid, and palmitic acid (Padmaja 2009). We have previously shown that sweet potatoes could be used to form an alternative low cost medium for cultivation of *Lactobacillus* (Hayek et al. 2013). However, the effect of sweet potatoes on the enzymatic activity of *Lactobacillus* was not determined. Therefore, the objective of this study was to study the enzymatic activity of *Lactobacillus* growing in a sweet potato-based medium.

Materials and Methods

Media preparation. Sweet potato medium (SPM) was prepared according to the method previously developed in our laboratory (Hayek et al. 2013). MRS was prepared by dissolving 55 g lactobacilli MRS and 1 g L-cysteine in 1 L deionized distilled water. SPM and MRS were then autoclaved at 121 °C for 15 min, cooled down, and stored at 4 °C to be used within 3 days.

Bacterial enumeration. Bacterial populations were determined by serial dilutions in 9 mL 0.1 % peptone water solution, 100 μL of appropriate dilutions were plated

Table 1 Lactobacillus strains, original sources, and final bacterial population after 16 h of incubation at 37 °C

Lactobacilli strain	Original source	Bacterial population log CFU/mL*	
		MRS	SPM
L. plantarum 299v	Human patient	10.73 ± 0.42^a	11.08 ± 0.43^a
L. acidophilus SD16	Commercial source	10.66 ± 0.41^a	10.98 ± 0.44^a
L. acidophilus EF7	Commercial source	10.96 ± 0.48^a	11.03 ± 0.56^a
L. rhamnosus GG B103	Child fecal isolate	11.00 ± 0.61^a	10.81 ± 0.61^a
L. delbrueckii subsp. bulgaricus SR35	Yogurt culture	10.88 ± 0.44^a	10.99 ± 0.14^a
L. reuteri CF2-7F	Child fecal isolate	11.14 ± 0.29^a	11.03 ± 0.61^a
L. reuteri SD2112	Mother's milk	11.06 ± 0.57^a	10.94 ± 0.41^a
Average		10.92 ± 0.55^a	10.98 ± 0.49^a

Data points are the average of 3 replicates with standard error
*Data points with different lower case letters in the same row are significantly ($p < 0.05$) different

onto MRS agar plates then incubated at 37 °C for 48 h. Plates with colonies ranging between 30 and 300 were considered for counting, and bacterial populations were expressed as log CFU/mL.

Culture conditions. Lactobacillus strains (Table 1) were activated by transferring 100 µL of stock culture to 10 mL MRS broth, incubated at 37 °C for 24 h, and stored at 4 °C. Prior to each experimental replication, bacterial strains were streaked on MRS agar and incubated for 48 h at 37 °C. *Lactobacillus* strains were individually subcultured twice in batches of 10 mL SPM broth and MRS broth separately for 24 h at 37 °C. Batches of 80 mL SPM and MRS media in 250 mL bottles were inoculated with 3 % v/v individual precultured *Lactobacillus* strains and incubated at 37 °C for 16 h after which final bacterial populations were determined.

Enzyme samples preparation. After incubation, cultures were divided into two portions at 40 mL each and cells were harvested by centrifugation at $7800 \times g$ for 10 min at 4 °C. Portions used for α-glucosidase and β-glucosidase were washed twice with 0.5 M sodium phosphate buffer (pH 6.0) and suspended in 1 mL of the same buffer. Portions used for phytase and acid phosphate were washed with 50 mM Tris–HCl (pH 6.5) and suspended in 1 mL 50 mM sodium acetate–acetic acid (pH 5.5). Suspended cells were maintained in Eppendorf tubes containing 0.1 mm glass beads and treated with a mini-Beadbeater-8 (Biospec Products, Bartlesville, OK, USA) for a total of 3 min in order to disrupt the cells and release the enzymes. Once per minute, samples were rested for 15 s in an ice bath to avoid overheating. Samples were then centrifuged ($12,000 \times g$ for 20 min) and supernatants were used for enzyme assay analysis of α-glucosidase, acid phosphatase, and phytase. Disrupted cells from the first portion were suspended in a minimum amount of sodium phosphate buffer and used for enzyme assay analysis of β-glucosidase.

Determination of α-/β-glucosidases. α-Glucosidase and β-glucosidase were determined by monitoring the rate of hydrolysis of p-nitrophenyl-α-D-glucopyranoside (α-pNPG) and p-nitrophenyl β-D-glucopyranoside (β-pNPG) substrates, respectively (Mahajan et al. 2010). For α-glucosidase, 1 mL of 10 mM α-pNPG was added to 0.5 mL of each sample and samples were then transferred to a water bath at 37 °C for 20 min. For β-glucosidase, 1 mL of 10 mM β-pNPG was added to 0.5 mL of each sample and samples were transferred into a water bath at 37 °C for 20 min. All reactions were stopped by adding 2.5 mL of 0.5 M Na_2CO_3. The released yellow p-nitrophenol was then determined by measuring the optical density at 420 nm. One unit of α-glucosidase or β-glucosidase (Glu U/mL) was defined as 1.0 μM of p-nitrophenol liberated per minute under assay conditions.

Determination of acid phosphatase and phytase. Acid phosphatase was determined by monitoring the rate of hydrolysis of p-nitrophenyl phosphate (p-NPP), and phytase was determined by measuring the amount of liberated inorganic phosphate (Pi) from sodium phytate (Haros et al. 2008). For acid phosphatase, 250 μL of 0.1 M sodium acetate buffer (pH 5.5) containing 5 mM p-NPP was mixed with 250 μL of enzyme sample. After incubation at 50 °C for 30 min in a water bath, the reaction was stopped by adding 500 μL of 1.0 M NaOH and the released p-nitrophenol was measured at 420 nm. For phytase, 400 μL of 0.1 M sodium acetate (pH 5.5) containing 1.2 mM sodium phytate was mixed with 200 μL of enzyme sample. After incubation for 30 min at 50 °C in a water bath, the reaction was stopped by adding 100 μL of 20 % trichloroacetic acid solution. An aliquot was analyzed to determine the liberated Pi by ammonium molybdate method, at 420 nm (Tanner and Barnett 1986). One unit of acid phosphatase or phytase (Ph U/mL) was defined as 1.0 μM of p-nitrophenol or 1.0 μM of Pi liberated per minute, respectively, under assay conditions.

Statistical analysis. Each test was conducted three times in randomized block design. R-Project for Statistical Computing version R-2.15.2 (www.r-project.org) was used to determine significant differences in the growth and enzymatic activity among *Lactobacillus* strains growing in SPM and MRS using one-way ANOVA (analysis of variance) with a significance level of $p < 0.05$.

Results and Discussion

Lactobacillus strains continued to grow at similar growth rates in SPM and MRS showing no significant ($p > 0.05$) differences in final bacterial populations after 16 h of incubation (Table 1). Table 2 shows α-glucosidase and β-glucosidase activity of *Lactobacillus* strains grown in SPM and MRS. α-Glucosidase activity ranged between 5.88 ± 0.58 and 56.93 ± 3.16 Glu U/mL. β-Glucosidase activity ranged between 2.61 ± 0.40 and 36.04 ± 3.16 Glu U/mL. Strains of *L. reuteri* and *L. delbrueckii* subsp. *bulgaricus* SR35 grown in SPM and MRS showed the highest α-glucosidase and β-glucosidase, respectively. *L. plantarum 299v* showed the lowest

Table 2 α-Glucosidase and β-glucosidase activities of *Lactobacillus* strains in SPM and MRS after 16 h of incubation at 37 °C

Lactobacilli strain	α-Glucosidase (Glu U/mL)*		β-Glucosidase (Glu U/mL)*	
	MRS	SPM	MRS	SPM
L. plantarum 299v	6.30 ± 0.80^{aD}	5.88 ± 0.58^{aD}	2.61 ± 0.40^{aD}	2.63 ± 0.51^{aD}
L. acidophilus SD16	7.72 ± 0.55^{aD}	8.54 ± 0.89^{aBC}	4.60 ± 0.77^{aCD}	2.66 ± 0.44^{bD}
L. acidophilus EF7	9.45 ± 0.73^{aCD}	9.63 ± 0.65^{aBC}	9.30 ± 0.57^{aC}	6.71 ± 0.79^{bC}
L. rhamnosus GG B103	10.50 ± 0.61^{aCD}	12.03 ± 0.56^{aB}	17.76 ± 1.04^{aB}	15.71 ± 1.76^{aB}
L. delbrueckii subsp. bulgaricus SR35	12.45 ± 0.79^{aC}	12.63 ± 0.86^{aB}	35.45 ± 2.70^{aA}	36.04 ± 3.16^{aA}
L. reuteri CF2-7F	56.93 ± 3.16^{aA}	31.87 ± 2.52^{bA}	4.72 ± 0.95^{bCD}	7.90 ± 0.76^{aC}
L. reuteri SD2112	47.30 ± 3.66^{aAB}	29.05 ± 2.14^{bA}	2.82 ± 0.38^{bD}	5.47 ± 0.96^{aCD}

Data points are the average of 3 replicates with standard error
*Data points with different lower case letters in the same row for the same enzyme are significantly ($p < 0.05$) different. Data points with different upper case letters in the same column are significantly ($p < 0.05$) different

α-glucosidase and β-glucosidase in both SPM and MRS. Strains of *L. acidophilus* and *L. rhamnosus* GG B103 growing in SPM showed higher α-glucosidase activity and lower β-glucosidase activity compared to MRS. *L. reuteri* showed lower α-glucosidase activity and higher β-glucosidase activity in SPM compared to MRS. *L. plantarum* 299v and *L. delbrueckii* subsp. *bulgaricus* SR35 showed similar α-glucosidase and β-glucosidase activity in both SPM and MRS. On average, lower α-glucosidase and higher β-glucosidase were observed in SPM compared to MRS.

Lactobacillus strains growing in SPM and MRS showed a wide range of differences in α-glucosidase and β-glucosidase. Enzymatic activity differences were observed between strains of the same species and for the same strain growing in different media. *These results agree with others finding that* α-glucosidase and β-glucosidase activity of *Lactobacillus* could be affected by media composition and could also vary among strains of the same species (Di Cagno et al. 2010; Mahajan et al. 2010; Otieno et al. 2005). Amounts and types of nutrients in culture media including carbohydrates, proteins, and minerals could affect α-glucosidase and β-glucosidase activity of *Lactobacillus* (Di Cagno et al. 2010; Mahajan et al. 2010; Otieno et al. 2005). In contrast, strains of *L. reuteri* and *L. delbrueckii* subsp. *bulgaricus* SR35 should be given more attention due to their exhibiting the highest α-glucosidase and β-glucosidase activity, respectively. Thus, SPM appeared to be convenient for *Lactobacillus* strains to grow and produce both α-glucosidase and β-glucosidase.

Table 3 shows acid phosphatase and phytase activity of *Lactobacillus* strains grown in SPM and MRS. Acid phosphatase activity in SPM and MRS ranged between 8.67 and 20.56 Ph U/mL. Phytase activity ranged between 0.16 ± 0.03 and 0.66 ± 0.14 Ph U/mL. Most *Lactobacillus* strains (*L. plantarum* 299v, *L. rhamnosus* GG B103, *L. reuteri* CF2-2F, and *L. reuteri* SD2112) showed higher acid phosphatase activity in SPM compared to MRS. With regard to phytase, all

Table 3 Acid phosphatase and phytase activities of *Lactobacillus* strains in SPM and MRS after 16 h of incubation at 37 °C

Lactobacilli strain	Acid phosphatase (Ph U/mL)*		Phytase (Ph U/mL)*	
	MRS	SPM	MRS	SPM
L. plantarum 299v	9.44 ± 1.18^{aB}	11.83 ± 1.09^{aC}	0.34 ± 0.03^{bA}	0.46 ± 0.04^{aB}
L. acidophilus SD16	11.22 ± 0.57^{aB}	9.92 ± 0.78^{aC}	0.32 ± 0.02^{bA}	0.41 ± 0.04^{aB}
L. acidophilus EF7	8.67 ± 0.87^{aB}	8.68 ± 0.85^{aC}	0.32 ± 0.05^{bA}	0.40 ± 0.03^{aB}
L. rhamnosus GG B103	9.96 ± 1.07^{aB}	10.60 ± 0.55^{aC}	0.30 ± 0.03^{bA}	0.40 ± 0.02^{aB}
L. delbrueckii subsp. bulgaricus SR35	11.19 ± 0.95^{aB}	10.61 ± 0.84^{aC}	0.19 ± 0.02^{aB}	0.16 ± 0.03^{bC}
L. reuteri CF2-7F	15.09 ± 0.95^{aA}	15.84 ± 1.05^{aB}	0.37 ± 0.03^{bA}	0.66 ± 0.14^{aA}
L. reuteri SD2112	13.50 ± 0.84^{bA}	20.56 ± 1.49^{aA}	0.39 ± 0.05^{bA}	0.65 ± 0.11^{aA}

Data points are the average of 3 replicates with standard error
*Data points with different lower case letters in the same row for the same enzyme are significantly ($p < 0.05$) different. Data points with different upper case letters in the same column are significantly ($p < 0.05$) different

Lactobacillus strains showed higher phytase activity in SPM except *L. delbrueckii* subsp. *bulgaricus* SR35. On average, acid phosphate and phytase activity of the tested *Lactobacillus* strains were higher in SPM than MRS. Strains of *L. reuteri* showed significantly ($p > 0.05$) the highest acid phosphatase and phytase activity in both SPM and MRS compared to other strains. A previous study also showed that *L. reuteri* could produce higher phytase compared to other lactobacilli species (Palacios et al. 2005). However, phytase activity in the tested strains seems to be low compared to acid phosphatase. Other studies have also shown that the phytase activity of *Lactobacillus* is generally low compared to other bacterial genera (Palacios et al. 2005; De Angelis et al. 2003). Thus, phytase does not seem to be common in *Lactobacillus* strains and as a result, *Lactobacillus* could express better acid phosphatase and phytase activity growing in SPM compared to MRS.

Conclusion

Our results demonstrate that *Lactobacillus* can grow and express their enzymatic activity in SPM. On average, the tested *Lactobacillus* strains showed higher β-glucosidase, acid phosphatase, and phytase but lower α-glucosidase in SPM compared to MRS. Differences in the enzymatic activity of *Lactobacillus* strains between SPM and MRS suggested that the medium composition has a significant effect on the enzymatic activity of *Lactobacillus*. *L. reuteri* (CF2-7F and SD2112) showed the highest α-glucosidase, acid phosphates, and phytase whereas *L. delbrueckii* subsp. *bulgaricus* SR35 showed the highest β-glucosidase activity. These strains should be given more attention in probiotic and food fermentation applications. Therefore, SPM could be a suitable medium to produce enhanced levels of β-glucosidase, acid phosphatase, and phytase activity of *Lactobacillus* and could also enhance other bioactivity.

Acknowledgments This work was made possible by grant number NC.X-267-5-12-170-1 from the National Institute of Food and Agriculture and its contents are solely the responsibility of the authors and do not necessarily represent the official view of the National Institute of Food and Agriculture. Part of this work was published in the journal of *British Microbiology Research Journal*, 2013, 4(5), 509–522, and part was presented at North Carolina Branch of the American Society for Microbiology. 2013 Annual Meeting, October 26th, East Carolina University, Greenville, NC.

References

De Angelis, M., Gallo, G., Corbo, M. R., McSweeney, P. L., Faccia, M., Giovine, M., et al. (2003). Phytase activity in sourdough lactic acid bacteria: Purification and characterization of a phytase from *Lactobacillus sanfranciscensis* CB1. *International Journal of Food Microbiology, 87*(3), 259–270.

Di Cagno, R., Mazzacane, F., Rizzello, C. G., Vincentini, O., Silano, M., Giuliani, G., et al. (2010). Synthesis of isoflavone aglycones and equol in soy milks fermented by food-related lactic acid bacteria and their effect on human intestinal Caco-2 cells. *Journal of Agricultural and Food Chemistry, 58*(19), 10338–10346.

Haros, M., Bielecka, M., Honke, J., & Sanz, Y. (2008). Phytate-degrading activity in lactic acid bacteria. *Polish Journal of Food and Nutrition Sciences, 58*(1), 33–40.

Hayek, S. A., & Ibrahim, S. A. (2013). Current limitations and challenges with lactic acid bacteria: A review. *Food and Nutrition Sciences, 4*(11A), 73–87.

Hayek, S. A., Shahbazi, A., Awaisheh, S. S., Shah, N. P., & Ibrahim, S. A. (2013). Sweet potatoes as a basic component in developing a medium for the cultivation of lactobacilli. *Bioscience, Biotechnology, and Biochemistry, 77*(11), 1–7.

Krasikov, V. V., Karelov, D. V., & Firsov, L. M. (2001). α-Glucosidases. *Biochemistry (Moscow), 66*(3), 267–281.

López-González, A. A., Grases, F., Roca, P., Mari, B., Vicente-Herrero, M. T., & Costa-Bauzá, A. (2008). Phytate (*myo*-inositol hexaphosphate) and risk factors for osteoporosis. *Journal of Medicinal Food, 11*(4), 747–752.

Mahajan, P. M., Desai, K. M., & Lele, S. S. (2010). Production of cell membrane-bound α-and β-glucosidase by *Lactobacillus acidophilus*. *Food and Bioprocess Technology, 5*, 706–718.

Otieno, D. O., Ashton, J. F., & Shah, N. P. (2005). Stability of β-glucosidase activity produced by *Bifidobacterium* and *Lactobacillus* spp. in fermented soymilk during processing and storage. *Journal of Food Science, 70*(4), M236–M241.

Padmaja, G. (2009). Uses and nutritional data of sweetpotato. In G. Loebenstein & G. Thottappilly (Eds.), *The sweetpotato* (Vol. 1, pp. 189–234). New York: Springer.

Palacios, M. C., Haros, M., Rosell, C. M., & Sanz, Y. (2005). Characterization of an acid phosphatase from *Lactobacillus pentosus*: Regulation and biochemical properties. *Journal of Applied Microbiology, 98*(1), 229–237.

Pfeiler, E. A., & Klaenhammer, T. R. (2007). The genomics of lactic acid bacteria. *Trends in Microbiology, 15*(12), 546–553. doi:10.1016/j.tim.2007.09.010.

Sestelo, A. B. F., Poza, M., & Villa, T. G. (2004). β-Glucosidase activity in a *Lactobacillus plantarum* wine strain. *World Journal of Microbiology & Biotechnology, 20*, 633–637.

Song, D., Ibrahim, S., & Hayek, S. (2012). Recent application of probiotics in food and agricultural science. In E. C. Rigobelo (Ed.), *Probiotics* (1st ed., Vol. 10, pp. 1–34). Manhattan, NY: InTech.

Tanner, J. T., & Barnett, S. A. (1986). Methods of analysis of infant formula: Food and drug administration and infant formula council collaborative study, phase III. *Journal of Association of Official Analytical Chemists, 69*, 777–785.

Zotta, T., Ricciardi, A., & Parente, E. (2007). Enzymatic activities of lactic acid bacteria isolated from Cornetto di Matera sourdoughs. *International Journal of Food Microbiology, 115*, 165–172.

Effect of Metal Ions on the Enzymatic Activity of *Lactobacillus reuteri* Growing in a Sweet Potato Medium

Saeed A. Hayek and Salam A. Ibrahim

Abstract The objective of this study was to determine the effect of metal ions on α-glucosidase, β-glucosidase, acid phosphatase, and phytase activity of *L. reuteri*. In the control group, *L. reuteri* MF14-C, MM2-3, SD2112, and DSM20016 produced the highest α-glucosidase (40.06 ± 2.80 Glu U/mL), β-glucosidase (17.82 ± 1.45 Glu U/mL), acid phosphatase (20.55 ± 0.74 Ph U/mL), and phytase (0.90 ± 0.05 Ph U/mL) respectively. The addition of Mg^{2+} and Mn^{2+} led to an increase in α-glucosidase activity of *L. reuteri* MM2-3 by 23.46 and 20.77 Glu U/mL respectively. The β-glucosidase activity of MM7 and SD2112 increased in the presence of Ca^{2+} by 9.65 and 9.85 and Fe^{2+} by 11.4 and 9.62 Glu U/mL respectively. Acid phosphatase produced by *L. reuteri* CF2-7F and MM2-3 increased in the presence of Mg^{2+}, Ca^{2+}, and Mn^{2+} by 13.53, 6.17, and 10.01 and 5.54, 5.92, and 3.98 Ph U/mL respectively. Phytase produced by *L. reuteri* MM2-3 increased in the presence of Mg^{2+} and Mn^{2+} by 0.26 and 0.38 Ph U/mL respectively. On average, Mg^{2+} and Mn^{2+} followed by Ca^{2+} led to the highest enhancement of tested enzymes. The effect of metal ions on the enzymatic activity of *L. reuteri* was found to be strain dependent. Thus, the enzymatic activity of *L. reuteri* could be enhanced by the addition of metal ions, with the specific selection of metal ions being necessary to maximize the level of the target enzyme.

Introduction

Foods from plants such as fruits, vegetables, cereals, legumes, and nuts contain different types of indigestible fibers including oligosaccharides, cellulose, and phytate. Plants are essential parts of human diet and obviously of great importance

S.A. Hayek
Department of Energy and Environmental Systems, North Carolina Agricultural and Technical State University, Greensboro, NC, USA

S.A. Ibrahim (✉)
North Carolina Agricultural and Technical State University, Greensboro, NC, USA
e-mail: ibrah001@ncat.edu

to the food industry. However, plant processing generates a large percentage of byproducts and waste. Indigestible fibers form a major part of wasted plant materials. In addition, undigested fiber in human diets could lead to health and nutritional problems. Thus, hydrolysis of such complex carbohydrates is getting more attention. *Lactobacillus* spp. is among the most important species in the food industry. *Lactobacillus* strains that can possess α-glucosidase, β-glucosidase, acid phosphatase, and phytase could help break down complex carbohydrates and also improve nutrient availability (Mahajan et al. 2010; Zotta et al. 2007). α-glucosidase (α-D-glucoside glucohydrolase, EC 3.2.1.20) is responsible for hydrolyzing glycosidic bonds in oligosaccharides (starch, disaccharides, and glycogen) and for releasing α-glucose. β-glucosidase (β-D-glucoside glucohydrolase, EC 3.2.1.21) can hydrolyze all four β-linked glucose dimmers in cellulose to produce glucose monomers. Acid phosphatase (orthophosphoric monoester phosphohydrolase, EC. 3.1.3.2) and phytase (*myo*-inositol hexakisphosphate 6-phosphohydrolases; EC 3.1.3.26) can break down phytate and reduce its antinutritional properties (Palacios et al. 2005). The specificity of acid phosphatase and phytase partially overlap since acid phosphatase produced by microorganisms has phytase activity. Due to the importance of *Lactobacillus* in the food industry and in probiotic applications, the enhancement of the enzymatic activity of *Lactobacillus* has been gaining more attention.

The nutritional requirements of *Lactobacillus* have been established. However, controlling, optimizing, and maximizing the enzymatic activity of *Lactobacillus* can be affected by different nutrients (Hayek and Ibrahim 2013). Ingredients such as metal ions, sugars, protein, and Tweens could enhance the enzymatic activity of *Lactobacillus* (Tham et al. 2010; Ibrahim et al. 2010; Hayek and Ibrahim 2013). The effect of metal ions on the enzymatic activity of *Lactobacillus* has been studied widely. For example, α-glucosidase activity was stimulated by Mn^{2+} and Mg^{2+} (Hendriksz and Gissen 2011). The addition of 10 mM Mn^{2+} caused a significant enhancement in β-glucosidase activity (Jeng et al. 2011). The enzymatic activity of acid phosphatase was enhanced by Ca^{2+} and Mg^{2+} with a greater effect due to Ca^{2+} (Tham et al. 2010). *Lactobacillus reuteri* is a probiotic species that inhabits the gastrointestinal tract of humans and animals (Casas and Dobrogosz 2000). *L. reuteri* exhibit high activity of α-galactosidase and β-galactosidase (Ibrahim et al. 2010), α-glucosidase (Kralj et al. 2005), and β-glucosidase (Otieno et al. 2005). Strains of *L. reuteri* also showed higher phytate degrading activity than other *Lactobacillus* spp., producing both phytase and acid phosphatase (Palacios et al. 2007). We have also shown that *L. reuteri* produce higher α-glucosidase, acid phosphatase, and phytase than other *Lactobacillus* spp. (Hayek et al. 2013b). Developing a means to enhance the enzymatic activity of *L. reuteri* could help to hydrolyze complex carbohydrates and to solve variety of digestive problems. Therefore, the objective of this study was to determine the effect of metal ions on the enzymatic activity of *L. reuteri* growing in a sweet potato-based medium.

Materials and Methods

Media and culture preparation. Sweet potato medium (SPM) was prepared according to the method that was previously developed in our laboratory (Hayek et al. 2013a). Strains of *L. reuteri* (Table 1) were activated in SPM broth by transferring 100 μL of stock culture to 10 mL SPM broth, incubated at 37 °C for 24 h, and stored at 4 °C. Prior to each experimental replication, bacterial strains were streaked on SPM agar and incubated for 48 h at 37 °C. One isolated colony was transferred to 10 mL SPM broth and incubated at 37 °C for next day use.

Culture conditions. To study the effect of metal ions on the enzymatic activity of *Lactobacillus*, 10 mM of $FeSO_4 \cdot 4H_2O$, $MgSO_4 \cdot 7H_2O$, K_2SO_4, or Na_2SO_4, and 5 mM of $MnSO_4 \cdot 4H_2O$ or $CaSO_4 \cdot 7H_2O$ were used. Metal ions were added separately to batches of 60 mL SPM broth in 250 mL bottles. Batches of 60 mL SPM without metal ions served as control. Samples were inoculated with a 3 % v/v precultured individual strain of *L. reuteri* and incubated at 37 °C for 16 h. Cultures were then divided into two portions of 30 mL each and cells were harvested by centrifugation at $7800 \times g$ for 10 min at 4 °C.

Preparation of enzyme samples. Portions of cultures used for α-glucosidase and β-glucosidase were washed twice with 0.5 M sodium phosphate buffer (pH 6.0) and suspended in 1 mL of the same buffer. Portions used for acid phosphatase and phytase were washed with 50 mM Tris–HCl (pH 6.5) and suspended in 1 mL of 50 mM sodium acetate-acetic acid (pH 5.5). Suspended cells were maintained in Eppendorf tubes containing 0.1 mm glass beads and treated with a Mini-BeadBeater-16 (Biospec Products, Bartlesville, OK, USA) for a total of 3 min. Once per minute, samples were rested for 15 s in ice bath to avoid overheating. Samples were then centrifuged ($12,000 \times g$ for 20 min) and supernatants were used for enzyme assay analysis of α-glucosidase, acid phosphatase, and phytase. Disrupted cells from the first portion were suspended in a minimum amount of sodium phosphate buffer and used for enzyme assay analysis of β-glucosidase.

Determination of α-glucosidase and β-glucosidase. α-Glucosidase and β-glucosidase were determined by monitoring the rate of hydrolysis of *p*-nitrophenyl-α-D-glucopyranoside (α-*p*NPG) and *p*-nitrophenyl-β-D-glucopyranoside (β-*p*NPG) respectively (Mahajan et al. 2010). For α-glucosidase, 1 mL of 10 mM α-*p*NPG substrate was added to 0.5 mL of enzyme sample, and samples were then

Table 1 *Lactobacillus reuteri* strains and sources that were used in this study

L. reuteri	Source
MF14C	Mother fecal isolate
CF2-7F	Child fecal isolate
DSM20016	Mother's milk
SD2112	Mother's milk
MM7	Mother's milk
MM2-3	Mother's milk

transferred into a water bath at 37 °C for 20 min. For β-glucosidase, 1 mL of 10 mM β-ρNPG was added to 0.5 mL of enzyme sample, and samples were then transferred to a water bath at 37 °C for 20 min. All reactions were stopped by adding 2.5 mL of 0.5 M Na_2CO_3. The released yellow ρ-nitrophenol was determined by measuring the optical density at 420 nm. One unit of α-glucosidase or β-glucosidase (Glu U/mL) was defined as 1.0 µM of ρ-nitrophenol liberated per minute under the assay conditions.

Determination of acid phosphatase and phytase. Acid phosphatase was determined by monitoring the rate of hydrolysis of ρ-nitrophenyl phosphate (ρ-NPP) and phytase was determined by measuring the amount of liberated inorganic phosphate (Pi) from sodium phytate (Haros et al. 2008). For acid phosphatase, 250 µL of 0.1 M sodium acetate buffer (pH 5.5) containing 5 mM ρ-NPP was mixed with 250 µL of enzyme sample. After incubation at 50 °C for 30 min in a water bath, the reaction was stopped by adding 0.5 mL of 1.0 M NaOH and the released ρ-nitrophenol was measured at 420 nm. For phytase, 400 µL of 0.1 M sodium acetate (pH 5.5) containing 1.2 mM sodium phytate was mixed with 250 µL of enzyme sample. After incubation for 30 min at 50 °C in a water bath, the reaction was stopped by adding 100 µL of 20 % trichloroacetic acid solution. An aliquot was analyzed to determine the liberated Pi by the ammonium molybdate method, at 420 nm (Tanner and Barnett 1986). One unit of acid phosphatase or phytase (Ph U/mL) was defined as 1.0 µM of ρ-nitrophenol or 1.0 µM of Pi liberated per minute respectively under the assay conditions.

Statistical analysis. Each experimental test was conducted three times in randomized block design. R-Project for Statistical Computing version R-2.15.2 (www.r-project.org) was used to determine significant differences in the effect of metal ions on the enzymatic activity of *L. reuteri* using one way and multiday ANOVA with a significance level of $p < 0.05$.

Results and Discussion

Induction of α-glucosidase by different metal ions. The tested strains of *L. reuteri* were able to survive and grow in the presence of metal ions. In the control, α-glucosidase activity ranged between 20.65 ± 1.70 and 40.06 ± 2.80 Glu U/mL (Table 2). The addition of metal ions resulted in an increase or decrease in the α-glucosidase activity of *L. reuteri*. The highest α-glucosidase activity (61.74 ± 3.09 and 58.31 ± 2.88 Glu U/mL) was obtained from MF14-C grown in the presence of Mg^{2+} and Mn^{2+} respectively. In general, Mg^{2+} and Mn^{2+} increased α-glucosidase activity whereas Fe^{2+} and K^+ caused a significant ($p < 0.05$) decrease. The addition of K^+ led to the highest negative impact on all *L. reuteri* strains. The addition of Ca^{2+} or Na^+ enhanced α-glucosidase activity of some strains including: DSM20016, MM7, and MM2-3 and decreased the activity of MF14C and SD2112. The α-glucosidase activity of *L. reuteri* CF2-7F was enhanced by the

Table 2 Effect of metal ions on α-glucosidase activity (Glu U/mL) produced by *L. reuteri*

L. reuteri	α-Glucosidase activity (Glu U/mL)						
	Control	Fe^{2+}	Ca^{2+}	Na$^+$	K$^+$	Mg^{2+}	Mn^{2+}
MF14C	40.06 ± 2.80^{bA}	25.90 ± 2.72^{cA}	26.32 ± 3.50^{cB}	38.38 ± 1.98^{bB}	13.71 ± 1.70^{dBC}	**61.74** ± 3.09^{aA}	58.31 ± 2.88^{aA}
CF2-7F	34.38 ± 1.36^{bB}	20.48 ± 1.75^{dB}	25.20 ± 1.83^{cB}	47.59 ± 1.86^{aA}	10.52 ± 1.30^{eD}	46.30 ± 2.64^{aBC}	32.83 ± 2.91^{bC}
DSM20016	31.80 ± 2.01^{cBC}	30.20 ± 2.27^{cA}	41.69 ± 2.89^{bA}	46.14 ± 2.66^{bA}	26.22 ± 1.09^{dA}	50.64 ± 1.30^{aB}	55.52 ± 4.68^{aA}
SD2112	29.34 ± 1.27^{cC}	19.54 ± 3.89^{deB}	20.57 ± 1.82^{dC}	34.58 ± 2.32^{abB}	15.59 ± 1.11^{deB}	35.51 ± 2.74^{abD}	40.55 ± 1.90^{aB}
MM7	25.32 ± 2.32^{bCD}	19.40 ± 1.75^{cB}	35.93 ± 3.62^{aA}	37.57 ± 4.39^{aB}	12.59 ± 1.83^{dCD}	28.13 ± 3.22^{abE}	35.68 ± 2.63^{aC}
MM2-3	20.65 ± 1.70^{cE}	18.28 ± 1.49^{cB}	24.79 ± 2.72^{bB}	24.98 ± 2.30^{bC}	14.79 ± 2.60^{eA}	44.11 ± 3.20^{aC}	41.42 ± 3.66^{aB}

^{a}Data points with different lower case letters in the same row are significantly ($p < 0.05$) different. Data points with different upper case letters in the same column are significantly ($p < 0.05$) different
Bold value: ($p < 0.001$)

addition of Na^{2+} and decreased by the addition of Ca^{2+}. These results indicate that the effect of Ca^{2+} and Na^+ on α-glucosidase activity could be strain dependent. The α-glucosidase activity of MM2-3 increased by 23.46 and 20.77 Glu U/mL in the presence of Mg^{2+} and Mn^{2+} respectively showing the highest increase in α-glucosidase. Since not all *L. reuteri* strains responded at the same level to the presence of metal ions, we conclude that the effect of metal ions on α-glucosidase activity of *L. reuteri* is strain dependent.

Induction of β-glucosidase by different metal ions. In the control group, β-glucosidase activity ranged between 6.94 ± 1.29 and 17.82 ± 1.45 Glu U/mL (Table 3). The additions of metal ions either enhanced or only produced a slight effect on β-glucosidase activity of *L. reuteri*. The highest β-glucosidase was observed from DSM20016 and MM2-3 in the presence of K^+ (26.22 ± 1.09 and 24.79 ± 3.24 Glu U/mL respectively). The addition of Ca^{2+} also caused a high increase in β-glucosidase activity of CF2-7F, DSM20016, and MM2-3, reaching 21.93 ± 2.05, 22.45 ± 1.70, and 22.70 ± 3.27 Glu U/mL respectively. However, the effect of metal ions on β-glucosidase activity produced by *L. reuteri* is not the same for all strains. For example, the addition of K^+ increased β-glucosidase produced by *L. reuteri* CF2-7F and *L. reuteri* SD2112 by 3.59 and 9.91 Glu U/mL respectively. On the other hand, β-glucosidase activity of *L. reuteri* MM7 increased by 11.60 Glu U/mL in the presence of Fe^{2+} and by only 1.90 Glu U/mL in the presence of Mn^{2+}. Therefore, the enhancement of β-glucosidase activity by metal ions was found to be strain dependent since the response of the strains to metal ions showed wide differences.

Induction of acid phosphatase by different metal ions. Table 4 shows acid phosphatase activity of *L. reuteri* in SPM with added metal ions. In the control samples, acid phosphatase activity ranged between 8.73 ± 1.11 and 20.56 ± 0.74 Ph U/mL. In the presence of metal ions, acid phosphatase ranged between 9.52 ± 1.14 (MM2-3 in the presence of Fe^{2+}) and 29.33 ± 2.36 (DSM20016 in the presence of Mg^{2+}) Ph U/mL. The presence of Mg^{2+} and Ca^{2+} showed a significant ($p < 0.05$) enhancement in acid phosphatase activity for all tested strains. Acid phosphatase activity of CF2-7F and MM2-3 increased by 13.53 (in the presence of Mg^{2+}) and 5.92 (in the presence of Ca^{2+}) Ph U/mL respectively. *L. reuteri* MM2-3 showed the highest enhancement with the addition of metal ions whereas *L. reuteri* DSM20016 showed the lowest enhancement with the addition of metal ions with regard to acid phosphatase activity. The addition of metal ions enhanced acid phosphatase activity of *L. reuteri* MM2-3 and DSM20016 by 39.7 % and 12.6 % respectively. The effect of metal ions on the activity of acid phosphatase was found to vary among *L. reuteri* strains. In general, the tested metal ions enhanced acid phosphatase activity of *L. reuteri*, with Mg^{2+}, Mn^{2+}, and Ca^{2+} showing the highest enhancement.

Induction of phytase by different metal ions. Table 5 shows phytase activity of *L. reuteri* in SPM with added metal ions. Phytase activity in control samples ranged between 0.51 ± 0.04 and 0.90 ± 0.05 Ph U/mL for strains MM2-3 and DSM20016 respectively. In the presence of metal ions, phytase ranged between 0.45 ± 0.04

Table 3 Effect of metal ions on β-glucosidase activity (Glu U/mL) produced by *L. reuteri*

L. reuteri	Control	β-Glucosidase activity (Glu U/mL)					
		Fe^{2+}	Ca^{2+}	Na$^+$	K$^+$	Mg^{2+}	Mn^{2+}
MF14C	6.94 ± 1.29^{cC}	11.46 ± 1.66^{abC}	**9.66 ± 2.17^{bC}**	11.18 ± 1.60^{abC}	13.71 ± 1.70^{aBC}	10.15 ± 1.74^{bD}	12.30 ± 1.29^{abB}
CF2-7F	10.11 ± 1.58^{dB}	18.02 ± 2.21^{abAB}	21.93 ± 2.05^{aA}	16.86 ± 1.12^{bB}	10.52 ± 1.30^{dD}	13.60 ± 1.13^{cC}	16.71 ± 1.93^{aA}
DSM20016	12.04 ± 1.05^{eB}	21.05 ± 2.23^{bcA}	22.45 ± 1.70^{bA}	18.24 ± 1.75^{cdB}	**26.22 ± 1.09^{aA}**	20.57 ± 1.28^{bcAB}	16.87 ± 1.08^{dA}
SD2112	7.59 ± 1.20^{cC}	17.21 ± 1.94^{aAB}	17.44 ± 1.90^{aA}	10.46 ± 1.69^{bcC}	15.59 ± 1.11^{abB}	17.50 ± 1.72^{aBC}	13.28 ± 1.50^{bB}
MM7	7.92 ± 0.88^{dC}	19.32 ± 2.94^{aAB}	17.57 ± 1.70^{aA}	12.37 ± 1.75^{aA}	12.59 ± 1.83^{aA}	14.28 ± 1.60^{aA}	9.82 ± 1.68^{aA}
MM2-3	17.82 ± 1.45^{bA}	23.60 ± 2.27^{aA}	22.70 ± 3.27^{aA}	23.48 ± 3.21^{aA}	24.79 ± 2.60^{aA}	22.57 ± 3.24^{aA}	17.52 ± 2.06^{bA}

^{a}Data points with different lower case letters in the same row are significantly ($p < 0.05$) different. Data points with different upper case letters in the same column are significantly ($p < 0.05$) different
Bold value: ($p < 0.001$)

Table 4 Effect of metal ions on acid phosphatase activity (Ph U/mL) produced by *L. reuteri*

L. reuteri	Control	Acid phosphatase activity (Ph U/mL)					
		Fe^{2+}	Ca^{2+}	Na$^+$	K$^+$	Mg^{2+}	Mn^{2+}
MF14C	13.59 ± 1.51^{bB}	14.51 ± 1.24^{bB}	19.14 ± 1.51^{aB}	13.49 ± 0.98^{bB}	13.74 ± 0.74^{bC}	22.24 ± 2.49^{aB}	14.06 ± 0.90^{bD}
CF2-7F	14.28 ± 1.2^{cB}	13.44 ± 1.11^{cB}	20.45 ± 2.16^{abB}	14.88 ± 1.57^{cB}	18.69 ± 1.16^{bB}	23.25 ± 2.38^{aB}	23.46 ± 2.07^{aAB}
DSM20016	18.69 ± 1.15^{bcA}	15.61 ± 2.40^{cB}	21.41 ± 2.01^{bB}	20.00 ± 1.27^{bA}	20.96 ± 2.19^{bB}	**29.33 ± 2.36^{aA}**	18.91 ± 1.95^{bcBC}
SD2112	20.55 ± 0.74^{bA}	19.73 ± 0.36^{bA}	24.78 ± 0.91^{aA}	20.50 ± 0.95^{bA}	26.64 ± 1.39^{aA}	27.81 ± 2.43^{aAB}	24.29 ± 2.54^{aA}
MM7	12.60 ± 1.63^{cB}	14.65 ± 0.98^{abB}	18.43 ± 2.66^{aB}	13.75 ± 2.37^{bcB}	13.74 ± 1.20^{bcC}	14.88 ± 1.31^{abC}	17.27 ± 2.53^{aC}
MM2-3	8.73 ± 1.11^{cC}	**9.52 ± 1.14^{bcC}**	14.65 ± 1.21^{aC}	11.91 ± 1.41^{abB}	10.12 ± 1.31^{bcD}	14.24 ± 1.49^{aC}	12.71 ± 1.37^{abD}

^{a}Data points with different lower case letters in the same row are significantly ($p < 0.05$) different. Data points with different upper case letters in the same column are significantly ($p < 0.05$) different

Bold value: ($p < 0.001$)

Table 5 Effect of metal ions on phytase activity (Ph U/mL) produced by *L. reuteri*

L. reuteri	Control	Phytase Fe^{2+}	Ca^{2+}	Na$^+$	K$^+$	Mg^{2+}	Mn^{2+}
MF14C	0.72 ± 0.08^{bcBC}	0.63 ± 0.06^{cA}	0.76 ± 0.11^{abcBC}	0.64 ± 0.05^{cAB}	0.93 ± 0.08^{aA}	0.81 ± 0.05^{abA}	0.87 ± 0.05^{abA}
CF2-7F	0.67 ± 0.05^{bcC}	0.62 ± 0.07^{bcA}	0.78 ± 0.08^{bB}	0.55 ± 0.04^{cB}	0.69 ± 0.05^{bBC}	0.81 ± 0.06^{bA}	0.95 ± 0.06^{aA}
DSM20016	0.90 ± 0.05^{abA}	0.67 ± 0.04^{cA}	**1.16 ± 0.20**^{aA}	0.46 ± 0.15^{cBC}	0.61 ± 0.23^{bcBC}	0.73 ± 0.19^{bAB}	0.81 ± 0.17^{abA}
SD2112	0.81 ± 0.06^{bAB}	0.64 ± 0.05^{cA}	0.99 ± 0.06^{aA}	0.68 ± 0.03^{bcA}	0.63 ± 0.04^{cC}	0.66 ± 0.08^{bcB}	1.03 ± 0.06^{aA}
MM7	0.68 ± 0.04^{bcC}	0.61 ± 0.06^{cA}	0.78 ± 0.06^{abB}	0.59 ± 0.09^{cB}	0.76 ± 0.06^{abB}	0.82 ± 0.04^{aA}	0.91 ± 0.14^{aA}
MM2-3	0.51 ± 0.04^{cD}	0.47 ± 0.04^{cA}	0.65 ± 0.04^{bC}	**0.45 ± 0.03**^{cC}	0.63 ± 0.11^{bBC}	0.77 ± 0.05^{bAB}	0.89 ± 0.03^{aA}

^{a}Data points with different lower case letters in the same row are significantly ($p < 0.05$) different. Data points with different upper case letters in the same column are significantly ($p < 0.05$) different
Bold value: $^bp < 0.001$

(MM2-3 in presence of Na^+) and 1.16 ± 0.20 (DSM20016 in presence of Ca^{2+}) Ph U/mL. The presence of Ca^{2+} or Mn^{2+} produces a significant ($p < 0.05$) enhancement in phytase activity for most tested strains. The presence of Fe^{2+} or Na^+ resulted in a significant ($p < 0.05$) reduction in phytase activity for most strains. The addition of K^+ or Mg^{2+} enhanced phytase activity of MF14C, MM7, and MM2-3 but showed a reduction or no effect on other strains. Some strains such as DSM20016 did not show any significant enhancement in phytase activity in the presence of metal ions. In contrast, the level of phytase activity of *L. reuteri* DSM20016 was reduced by several metal ions, Fe^{2+}, Na^+, K^+, and Mg^{2+}. Phytase activity of *L. reuteri* MF14C and MM7 was enhanced by all metal ions except Fe^{2+} and Na^{2+}. Thus, the effect of metal ions on phytase activity of *L. reuteri* differed widely in relation to different strains and different metal ions.

Phytase in *L. reuteri* was found to be low compared to that in other tested enzymes even with the enhancement by metal ions. Palacios and others (2005) showed that *Lactobacillus* strains had higher activity against p-nitrophenyl phosphate than against phytate. Phytase does not appear to be common in *Lactobacillus* spp. Other studies have also shown that phytase activity of *Lactobacillus* is generally low compared to that of other bacterial genera (Palacios et al. 2005; De Angelis et al. 2003). Phytase and acid phosphatase are particular subgroups of phosphatases, however; phytase has a preference for phytate. The specificity of both acid phosphatase and phytase can partially overlap since acid phosphatase also has phytase activity.

Conclusion

In this study, we explored the effect of metal ions on the enzymatic activity of *L. reuteri*. The enzymatic activity of *L. reuteri* was enhanced by the addition of different metal ions. However, to obtain the highest activity of a specific enzyme, selection of combination of metal ion and strain might be required. Mn^{2+} and Mg^{2+} showed the highest enhancement for the enzymatic activity of *L. reuteri*. Thus, it is recommended that Mn^{2+} and Mg^{2+} be added to enhance the overall enzymatic activity of *L. reuteri*. *L. reuteri* DSM20016 needs to be given more attention since this strain showed high enzymatic activity for all tested enzymes and showed a better response to metal ions compared to other strains. However, more work is required to determine the optimum concentration or possible combination of metal ions in order to reach the highest enzymatic activity of *L. reuteri*.

Acknowledgments This work was made possible by grant number NC.X-267-5-12-170-1 from the National Institute of Food and Agriculture and its contents are solely the responsibility of the authors and do not necessarily represent the official view of the National Institute of Food and Agriculture. Part of this work was published in the *Journal of SpringerPlus*, 2013, 2(465) and part was presented at North Carolina Branch of the American Society for Microbiology. 2013 Annual Meeting, October 26th, East Carolina University, Greenville, NC. *Lactobacillus reuteri* strains were provided by BioGaia, Raleigh, NC.

References

Casas, I. A., & Dobrogosz, W. J. (2000). Validation of the probiotic concept: *Lactobacillus reuteri* confers broad-spectrum protection against disease in humans and animals. *Microbial Ecology in Health and Disease, 12*(4), 247–285.

De Angelis, M., Gallo, G., Corbo, M. R., McSweeney, P. L., Faccia, M., Giovine, M., et al. (2003). Phytase activity in sourdough lactic acid bacteria: Purification and characterization of a phytase from *Lactobacillus sanfranciscensis* CB1. *International Journal of Food Microbiology, 87*(3), 259–270.

Haros, M., Bielecka, M., Honke, J., & Sanz, Y. (2008). Phytate-degrading activity in lactic acid bacteria. *Polish Journal of Food and Nutrition Sciences, 58*(1), 33–40.

Hayek, S. A., & Ibrahim, S. A. (2013). Current limitations and challenges with lactic acid bacteria: A review. *Food and Nutrition Sciences, 4*(11A), 73–87.

Hayek, S. A., Shahbazi, A., Awaisheh, S. S., Shah, N. P., & Ibrahim, S. A. (2013a). Sweet potatoes as a basic component in developing a medium for the cultivation of lactobacilli. *Bioscience, Biotechnology, and Biochemistry, 77*(11), 2248–2254.

Hayek, S. A., Shahbazi, A., Worku, M., & Ibrahim, S. A. (2013b). Enzymatic activity of lactobacillus grown in a sweet potato base medium. *British Microbiology Research Journal, 4*(5), 509–522.

Hendriksz, C. J., & Gissen, P. (2011). Glycogen storage disease. *Paediatrics and Child Health, 21*(2), 84–89. doi:C10.228.140.163.100.435.340.

Ibrahim, S. A., Alazzeh, A. Y., Awaisheh, S. S., Song, D., Shahbazi, A., & AbuGhazaleh, A. A. (2010). Enhancement of α-and β-galactosidase activity in *Lactobacillus reuteri* by different metal ions. *Biological Trace Element Research, 136*(1), 106–116.

Jeng, W. Y., Wang, N. C., Lin, M. H., Lin, C. T., Liaw, Y. C., Chang, W. J., et al. (2011). Structural and functional analysis of three β-glucosidases from bacterium *Clostridium cellulovorans*, fungus *Trichoderma reesei* and termite *Neotermes koshunensis*. *Journal of Structural Biology, 173*, 46–56.

Kralj, S., Stripling, E., Sanders, P., van Geel-Schutten, G. H., & Dijkhuizen, L. (2005). Highly hydrolytic reuteransucrase from probiotic *Lactobacillus reuteri* strain ATCC 55730. *Applied and Environmental Microbiology, 71*(7), 3942–3950.

Mahajan, P. M., Desai, K. M., & Lele, S. S. (2010). Production of cell membrane-bound α-and β-glucosidase by *Lactobacillus acidophilus*. *Food and Bioprocess Technology, 5*, 706–718.

Otieno, D. O., Ashton, J. F., & Shah, N. P. (2005). Stability of β-glucosidase activity produced by *Bifidobacterium* and *Lactobacillus* spp. in fermented soymilk during processing and storage. *Journal of Food Science, 70*(4), M236–M241.

Palacios, M. C., Haros, M., Rosell, C. M., & Sanz, Y. (2005). Characterization of an acid phosphatase from *Lactobacillus pentosus*: Regulation and biochemical properties. *Journal of Applied Microbiology, 98*(1), 229–237.

Palacios, M. C., Haros, M., Sanz, Y., & Rosell, C. M. (2007). Selection of lactic acid bacteria with high phytate degrading activity for application in whole wheat breadmaking. *LWT Food Science and Technology, 41*, 82–92.

Tanner, J. T., & Barnett, S. A. (1986). Methods of analysis of infant formula: Food and drug administration and infant formula council collaborative study, phase III. *Journal-Association of Official Analytical Chemists, 69*, 777–785.

Tham, S., Chang, C., Huang, H., Lee, Y., Huang, T., & Chang, C. (2010). Biochemical characterization of an acid phosphatase from *Thermus thermophilus*. *Bioscience, Biotechnology, and Biochemistry, 74*, 727–735.

Zotta, T., Ricciardi, A., & Parente, E. (2007). Enzymatic activities of lactic acid bacteria isolated from Cornetto di Matera sourdoughs. *International Journal of Food Microbiology, 115*, 165–172.

Using Sweet Potatoes as a Basic Component to Develop a Medium for the Cultivation of Lactobacilli

Saeed A. Hayek, Abolghasem Shahbazi, and Salam A. Ibrahim

Abstract Sweet potatoes (*Ipomoea batatas*) were investigated as a basic component to develop a medium for the cultivation of lactobacilli. Extract from baked sweet potatoes was used to form a sweet potato medium (SPM) which was supplemented with 0, 4, or 8 g/L of each nitrogen source (beef extract, yeast extract, and proteose peptone #3) to develop SPM1, SPM2, and SPM3, respectively. The growth of *Lactobacillus* in SPM was compared to lactobacilli MRS. Low inoculums' levels (2–2.5 log CFU/mL) were used to investigate the suitability of SPM to support the growth of *Lactobacillus*. *Lactobacillus* strains were individually inoculated into batches of MRS, SPM1, SPM2, and SPM3 then incubated at 37 °C. The growth of *Lactobacillus* was monitored using turbidity (OD at 610 nm), bacterial population (log CFU/mL), and pH values. Our results showed no significant differences ($p < 0.05$) in the *maximum specific growth rates* (μ_{max}) of *Lactobacillus* strains growing in MRS, SPM2, and SPM3. After 24 h on incubation, *Lactobacillus* strains grown in SPM2, SPM3, and MRS reached averages of 10.59 ± 0.27, 10.72 ± 0.19, and 10.41 ± 0.35 log CFU/mL, respectively. Slower growth rates were observed in SPM1 with 1.57 ± 0.55 log CFU/mL less of bacterial populations than MRS. SPM2 and SPM3 maintained higher pH values throughout the incubation period compared to MRS. Therefore, these findings indicated that SPM2, containing 12 g/L of nitrogen sources, is suitable for the growth of *Lactobacillus*, and SPM2 could be used as an alternative low-cost medium.

S.A. Hayek • S.A. Ibrahim (✉)
North Carolina Agricultural and Technical State University, Greensboro, NC 27411-1064, USA
e-mail: sahayek@ncat.edu; ibhah001@ncat.edu

A. Shahbazi
Natural Resources and Environmental Design, North Carolina Agricultural and Technical State University, Greensboro, NC 27411, USA
e-mail: ash@ncat.edu

Introduction

Lactobacilli are important for the production of fermented products and for probiotic applications (Song et al. 2012). This group of bacteria has fastidious nutritional requirements that vary among species and even strains in the same specie (Hayek and Ibrahim 2013; Vera Pingitore et al. 2009; Wegkamp et al. 2010). Various nutrients including carbohydrates, different nitrogen sources, fatty acid esters, minerals, and vitamins are required for normal growth of *Lactobacillus* (Hébert et al. 2004; Wegkamp et al. 2010). Due to the fastidious nutritional requirement of *Lactobacillus*, formulating cultivation media for lactobacilli could face various limitations and challenges (Hayek and Ibrahim 2013). Thus, standard culturing media such as MRS and M17 are expensive, require specific preparation steps, and need a long incubation time (Vazquez et al. 2004; Horn et al. 2005; Hayek and Ibrahim 2013; Djeghri-Hocine et al. 2007). Cost and incubation time could be the most important issues with regard to industrial applications (Hayek and Ibrahim 2013). High cost is introduced mainly by the use of expensive nitrogen sources such as beef extract, yeast extract, and peptone (Altaf et al. 2007; Djeghri-Hocine et al. 2007; Vazquez et al. 2004; Horn et al. 2005). Thus, finding alternative low-cost ingredients may reduce the media cost and have economic impact on the food industry.

Food industry byproducts, agriculture products, and agriculture wastes, as low-cost materials and rich nutrient sources, have shown evidence to support the growth of *Lactobacillus* (Djeghri-Hocine et al. 2007; Vazquez et al. 2004; Altaf et al. 2007). Sweet potatoes (*Ipomoea batatas*) are an abundant agricultural product in the United States and the state of North Carolina is the leading producer of sweet potatoes accounting for 47 % of the total production (Lucier and Glaser 2011). Sweet potato is a rich source of carbohydrates, some amino acids, vitamins, minerals, and dietary fiber (Padmaja 2009). Sweet potato also contains other minor nutrients such as antioxidants, triglycerides, linoleic acid, and palmitic acid (Padmaja 2009). Oleic acid is an essential growth factor for *Lactobacillus* strains (Jenkins and Courtney 2003) whereas antioxidants may stimulate the growth of *Lactobacillus* (Duda-Chodak et al. 2008). Because of the nutritional profile of sweet potatoes, this tuber has a great potential to partially or fully replace the expensive ingredients in *Lactobacillus* media and thus lower the cost. Therefore, this study aimed to study sweet potatoes as basic component in developing a medium for the cultivation of *Lactobacillus*.

Materials and Methods

Microorganisms. *Lactobacillus* strains were activated individually by transferring 100 μL of stock cultures to 10 mL of MRS broth and incubating at 37 °C for 24 h. Strains of *Lactobacillus* were then stored at 4 °C. Individual bacterial strain was streaked onto MRS agar and incubated for 48 h at 37 °C. Prior to each experimental replicate, one isolated colony was transferred to 10 mL of MRS broth and incubated at 24 h for 37 °C.

Preparation of Sweet Potato Medium. Sweet potatoes (Covington cultivar) were obtained from Burch Farms in Faison NC, USA. For each experimental replication, 900 g of fresh sweet potatoes were baked in a conventional oven at 400 °C for 1 h, cooled down, peeled, and blended with deionized distilled water (DDW) at ratio 1:2. This solution was centrifuged at 7800 × g for 10 min using Thermo Scientific* Sorvall RC 6 Plus Centrifuge (Thermo Scientific Co., Asheville, NC) and the supernatant was collected to form sweet potato extract (SPE). SPE was used to form three SPMs (SPM1, SPM2, and SPM3; Table 1) at different concentrations of nitrogen sources. MRS was prepared by dissolving 55 g MRS broth and 1 g L-cysteine in 1 L of DDW. SPMs and MRS were autoclaved at 121 °C for 15 min, cooled, and stored at 4 °C.

Bacterial Enumeration. Bacterial populations were determined by plating onto MRS agar. In this procedure, samples were individually diluted into serial of 9 mL 0.1 % peptone water solution then 100 µL of appropriate dilutions were surface-plated onto triplicates of MRS agar plates and incubated at 37 °C for 48 h. Plates with colonies ranging between 25 and 250 were considered for colony counting at the end of incubation to determine the bacterial populations.

Culture Conditions. Lactobacillus strains were individually subcultured twice in batches of 10 mL SMP1 broth and 10 mL of MRS broth for 24 h at 37 °C. This step was required to allow *Lactobacillus* strains adopt to the SPM. Overnight grown strains were serially diluted in 0.1 % peptone water solution then 1 mL of 4–5 log CFU/mL dilutions were transferred to batches of 50 mL corresponding media, mixed thoroughly, and incubated at 37 °C. Initial bacterial populations were determined and bacterial growth was monitored by measuring the turbidity (optical density at 610 nm) at 2 h interval during 20 h using Thermo Scientific Genesys 10S UV-Vis spectrophotometer (Thermo Fisher Scientific Co., Madison, WI); bacterial

Table 1 Composition of SPM for 1 L of sweet potato extract

Component	Supplier	Components (g/L)		
		SPM1	SPM2	SPM3
Proteose peptone #3	Remel	0	4.0	8.0
Beef extract	Neogen Corporation	0	4.0	8.0
Yeast extract	Neogen Corporation	0	4.0	8.0
Sodium acetate (CH_3COONa)	Fisher Scientific	5.0	5.0	5.0
Tween 80	Fisher Scientific	1.0 mL	1.0 mL	1.0 mL
Disodium phosphate (Na_2HPO_4)	Fisher Scientific	2.0	2.0	2.0
Ammonium citrate ($NH_4C_6H_5O_7$)	Fisher Scientific	2.0	2.0	2.0
Magnesium sulfate ($MgSO_4 \cdot 7H_2O$)	Fisher Scientific	0.1	0.1	0.1
Manganese sulfate ($MnSO_4 \cdot 5H_2O$)	Fisher Scientific	0.05	0.05	0.05
L-Cysteine	Fisher Scientific	1.0	1.0	1.0

populations (log CFU/mL) at 8, 12, 16, and 24 h; and pH at 12, 24, and 48 h. The maximum specific growth rates per hour (μ_{max}) for the tested *Lactobacillus* strains were determined during exponential growth phase using optical density values.

Determination of pH Value and Buffering Capacity. The pH values were determined using a pH meter (Accumet basic, AB15/15+, Fisher Scientific, Pittsburgh. PA, USA). Buffering capacity (BC) was determined by placing 25 mL of the medium into a 100 mL beaker then initial pH value was measured using a calibrated pH meter then titration was carried out using 0.1 N HCl until pH 2.0 reached (Salaün et al. 2005). Buffering capacity was calculated mathematically as the amount of 0.1 N HCl (mL) added divided by the unit change in pH values.

Statistical Analysis. Each experimental test was conducted three times in a randomized block design. R-Project for Statistical Computing version R-2.15.2 (www.r-project.org) was used to determine significant differences between MRS and SPMs using one way ANOVA (analysis of variance) at a significance level of $p < 0.05$.

Results and Discussion

Bacterial Growth. Table 2 shows the maximum growth rates (μ_{max}) of *Lactobacillus* strains growing in MRS and SPMs. *Lactobacillus* strains continue to grow in SPM2, SPM3, and MRS at similar growth rate. The growth of *Lactobacillus* strains showed no significant ($p > 0.05$) differences in μ_{max} values among SPM2, SPM3, and MRS. The growth of *Lactobacillus* in SPM1 showed significantly ($p < 0.05$)

Table 2 Maximum specific growth for the tested *Lactobacillus* strains in MRS, SPM1, SPM2, and SPM3

Lactobacillus strains	Maximum specific growth rate (μ_{max})			
	MRS	SPM1	SPM2	SPM3
L. plantarum 299v	0.334 ± 0.009^a	0.263 ± 0.028^b	0.353 ± 0.021^a	0.350 ± 0.021^a
L. acidophilus EF7	0.360 ± 0.017^a	0.237 ± 0.024^b	0.379 ± 0.030^a	0.357 ± 0.017^a
L. acidophilus SD16	0.384 ± 0.014^a	0.238 ± 0.014^b	0.371 ± 0.015^a	0.391 ± 0.021^a
L. delbrueckii subsp. *bulgaricus* SD33	0.396 ± 0.026^a	0.283 ± 0.016^b	0.387 ± 0.011^a	0.406 ± 0.024^a
L. delbrueckii subsp. *bulgaricus* SR35	0.366 ± 0.013^a	0.251 ± 0.021^b	0.334 ± 0.041^{ab}	0.359 ± 0.015^a
L. rhamnosus GG B101	0.344 ± 0.023^a	0.284 ± 0.029^b	0.381 ± 0.017^a	0.364 ± 0.015^a
L. rhamnosus GG B103	0.362 ± 0.027^{ab}	0.280 ± 0.039^b	0.355 ± 0.011^{ab}	0.374 ± 0.019^a
L. reuteri CF2-2F	0.347 ± 0.017^a	0.269 ± 0.032^b	0.371 ± 0.022^a	0.373 ± 0.018^a
L. reuteri SD2112	0.388 ± 0.027^a	0.267 ± 0.024^b	0.351 ± 0.034^a	0.368 ± 0.011^a
L. reuteri DSM20016	0.368 ± 0.022^a	0.251 ± 0.016^b	0.361 ± 0.029^a	0.389 ± 0.026^a

Values with different lower case letter in the same row are significantly ($p < 0.05$) different

Fig. 1 Picture shows greenish cloud formation in SPM2 and MRS compared to fresh SPM2 and MRS due to the growth of *Lactobacillus* after 24 h of incubation at 37 °C

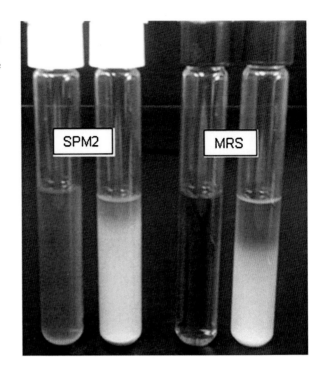

Table 3 Average of bacterial populations for *Lactobacillus* strains in MRS, SPM1, SPM2, and SPM3 at 8, 12, 16, and 2 h of incubation at 37 °C in log CFU/mL

Media	Incubation time (h)			
	8	12	16	24
	Bacterial population (log CFU/mL)			
MRS	4.25 ± 0.26^a	5.84 ± 0.36^a	8.25 ± 0.45^a	10.41 ± 0.35^a
SPM1	4.21 ± 0.29^a	5.17 ± 0.36^{ab}	7.12 ± 0.56^b	8.84 ± 0.82^b
SPM2	4.29 ± 0.25^a	5.95 ± 0.41^a	8.21 ± 0.31^a	10.59 ± 0.27^a
SPM3	4.34 ± 0.27^a	5.86 ± 0.40^a	8.27 ± 0.30^a	10.72 ± 0.19^a

Values with different lower case letters in the same column are significantly ($p < 0.05$) different

lower μ_{max} values than other media. The growth of *Lactobacillus* in SPM2 and MRS caused development of similar greenish cloud (Fig. 1). The average population of *Lactobacillus* strains throughout the incubation time showed no significant ($p > 0.05$) differences among MRS, SPM2, and SPM3 (Table 3). With regard to SPM1, the average bacterial population was slightly lower than MRS. However, after 24 h of incubation the average bacterial population in SPM1 was 1.57 ± 0.55 log CFU/mL less than MRS.

Results of bacterial populations and μ_{max} values agreed in that SPM2 and SPM3 could support similar level of growth compared to MRS. SPM2, containing 12 g/L of nitrogen sources compared to 25 g/L in MRS, showed similar growth rates and slightly higher bacterial population than MRS. On the other hand, the increased amount of nitrogen sources in SPM3 did not improve the growth rates of the tested *Lactobacillus* strains. This is supported by other findings, where an increase in amino acids and peptides beyond the essential requirements may not necessary result in increased growth (Lechiancole et al. 2002). Raw sweet potatoes contain about 1.57 % protein which will be increased to 2.01 % after baking (USDA 2012). Protein in sweet potato is the least affected nutrient by the cooking processes such as steaming, boiling, baking (Padmaja 2009). According to the current pricing of materials used in this study, formulating 1 L of media would cost US$4.46, 3.62, 2.37, and 1.1 for MRS, SPM3, SPM2, and SPM1, respectively. Therefore, SPM2 appears to be a suitable low-cost medium for normal growth of lactobacilli and could be used as an alternative to MRS. SPM1, without additional nitrogen sources, showed only 1.57 ± 0.55 log CFU/mL less than MRS after 24 h of incubation. The growth of *Lactobacillus* in the industrially used bulky media is lower than that in the standard media such as MRS (Fig. 1). Since sweet potatoes are the main ingredient in SPM1 and nitrogen source were excluded, SPM1 could be also used as food grade medium after slight modification (Sawatari et al. 2006). Thus, sweet potatoes, a rich source of several nutrients, could partially replace the expensive nitrogen sources in lactobacilli media and formula of SPM2 could be used as an alternative low-cost medium.

Acid Production and Buffering Capacity. Genus *Lactobacillus* belong to lactic acid bacteria group which is known to convert sugars to lactic acid. The acid production will result in a decrease in the medium pH which can be used as a growth indicator. Most *Lactobacillus* strains grow at optimum pH between 5.0 and 6.0, but they can also grow at a relatively low pH (4.4) (Von Wright and Axelsson 2011). The addition of buffering agents to the growth media is required to maintain the medium pH at optimum for normal growth. The initial pH values for the tested media ranged between 6.37 and 6.54. There was no difference in the pH values during the first 12 h of incubation among the media (Table 4).

SPM2 and SPM3 were able to maintain the pH above 4.4 after 24 h of incubation and around 4.4 after 48 h of incubation. The pH values in MRS dropped below 4.4 after 24 h and below 4.0 after 48 h of incubation. SPM2 and SPM3 showed almost

Table 4 Average pH values for MRS, SPM1, SPM2, and SPM3 due to the growth of *Lactobacillus*

Media	pH values		
	Incubation time (h)		
	12	24	48
MRS	5.989 ± 0.22^a	4.083 ± 0.18^c	3.75 ± 0.11^b
SPM1	5.992 ± 0.17^a	4.918 ± 0.38^a	4.43 ± 0.26^a
SPM2	5.862 ± 0.29^a	4.524 ± 0.22^b	4.34 ± 0.30^a
SPM3	5.893 ± 0.31^a	4.485 ± 0.21^b	4.33 ± 0.29^a

Data points are averages of three replicates of all tested *Lactobacillus* strains with standard error
Values with different lower case letter in the same column are significantly ($p < 0.05$) different

similar pH values throughout the incubation time. BC values for MRS, SPM1, SPM2, and SPM3 were 11.56 ± 0.51, 10.57 ± 0.44, 13.95 ± 0.52, and 14.73 ± 0.61, respectively. Higher BC in SPM2 and SPM3 may explain the ability of these media to maintain higher pH values throughout the incubation time compared to MRS. SPM1 also showed higher pH values compared to MRS; however, the growth of *Lactobacillus* in SPM1 was lower than other media which may contribute to lower acid production. Thus, SPM2 could better serve the growth of *Lactobacillus* than MRS, while maintaining higher pH values.

Conclusion

A new medium (SPM) based on the use of sweet potatoes was developed in our laboratory. Since the growth of *Lactobacillus* strains in SPM2 was similar to MRS, formula in SPM2 could be used as bulky low-cost medium in the food industry. Sweet potatoes could partially replace the expensive nitrogen sources in lactobacilli media and lower the cost. Using sweet potatoes to develop a new medium for lactobacilli could open the door to new applications for sweet potato in North Carolina and the United States. Future work in our laboratory will be conducted to formulate the final dehydrated product of SPM.

Acknowledgments This work was made possible by grant number NC.X-267-5-12-170-1 from the National Institute of Food and Agriculture and its contents are solely the responsibility of the authors and do not necessarily represent the official view of the National Institute of Food and Agriculture. The full paper of this work was published in the journal of *Bioscience, Biotechnology, and Biochemistry*, 2013, 77(11), 2248–2254, and part was presented at the 17th Biennial Research Symposium of the Association of 1890 Research Directors, Inc. (ARD) April, 2013.

References

Altaf, M. D., Naveena, B. J., & Reddy, G. (2007). Use of inexpensive nitrogen sources and starch for L (+) lactic acid production in anaerobic submerged fermentation. *Bioresource Technology, 98*(3), 498–503.

Djeghri-Hocine, B., Boukhemis, M., Zidoune, M. N., & Amrane, A. (2007). Evaluation of de-lipidated egg yolk and yeast autolysate as growth supplements for lactic acid bacteria culture. *International Journal of Dairy Technology, 60*(4), 292–296.

Duda-Chodak, A., Tarko, T., & Statek, M. (2008). The effect of antioxidants on *Lactobacillus casei* cultures. *ACTA Scientiarum Polonorum Technologia Alimentaria, 7*, 39–51.

Hayek, S. A., & Ibrahim, S. A. (2013). Current limitations and challenges with lactic acid bacteria: A review. *Food and Nutrition Sciences, 4*(11A), 73–87.

Hébert, E. M., Raya, R. R., & de Giori, G. S. (2004). Evaluation of minimal nutritional requirements of lactic acid bacteria used in functional foods. In J. M. Walker, J. F. Spencer, & A. L. Ragout de Spencer (Eds.), *Methods in biotechnology* (pp. 139–150). Totowa, NJ: Humana Press.

Horn, S. J., Aspmo, S. I., & Eijsink, V. G. H. (2005). Growth of *Lactobacillus plantarum* in media containing hydrolysates of fish viscera. *Journal of Applied Microbiology, 99*, 1082–1089. doi:10.1111/j.1365-2672.2005.02702.x.

Jenkins, J. K., & Courtney, P. D. (2003). *Lactobacillus* growth and membrane composition in the presence of linoleic or conjugated linoleic acid. *Canadian Journal of Microbiology, 49*(1), 51–57.

Lechiancole, T., Ricciardi, A., & Parente, E. (2002). Optimization of media and fermentation conditions for the growth of *Lactobacillus sakei*. *Annals of Microbiology, 52*(3), 257–274.

Lucier, G., & Glaser, L. (2011). *Vegetables and melons outlook*. USDA, A Report from the Economic Service VGS-343 (pp. 1–47).

Padmaja, G. (2009). Uses and nutritional data of sweetpotato. In G. Loebenstein & G. Thottappilly (Eds.), *The sweetpotato* (Vol. 1, pp. 189–234). Dordrecht, The Netherlands: Springer.

Salaün, F., Mietton, B., & Gaucheron, F. (2005). Buffering capacity of dairy products. *International Dairy Journal, 15*(2), 95–109.

Sawatari, Y., Hirano, T., & Yokota, A. (2006). Development of food grade media for the preparation of *Lactobacillus plantarum* starter culture. *The Journal of General and Applied Microbiology, 52*(6), 349–356.

Song, D., Ibrahim, S., & Hayek, S. (2012). Recent application of probiotics in food and agricultural science. In E. C. Rigobelo (Ed.), *Probiotics* (Vol. 10, pp. 1–34). Manhattan, NY: InTech.

USDA. (2012). *Nutrient data for 11508, sweet potato, cooked, baked in skin, without salt. Nutrient Data Laboratory*. Washington, DC: U.S. Department of Agriculture.

Vazquez, J. A., Gonzalez, M. P., & Murado, M. A. (2004). Peptones from autohydrolysed fish viscera for nisin and pediocin production. *Journal of Biotechnology, 112*(3), 299–311.

Vera Pingitore, E., Hebert, E. M., Sesma, F., & Nader-Macias, M. E. (2009). Influence of vitamins and osmolites on growth and bacteriocin production by *Lactobacillus salivarius* CRL 1328 in a chemically defined medium. *Canadian Journal of Microbiology, 55*(3), 304–310. doi:10.1139/w08-092.

Von Wright, A., & Axelsson, L. (2011). Lactic acid bacteria: An introduction. In S. Lahtinne, S. Salminen, A. Von Wright, & A. Ouwehand (Eds.), *Lactic acid bacteria: Microbiological and functional aspects* (pp. 1–17). London: CRC Press.

Wegkamp, A., Teusink, B., De Vos, W. M., & Smid, E. J. (2010). Development of a minimal growth medium for *Lactobacillus plantarum*. *Letters in Applied Microbiology, 50*(1), 57–64.

Impact of Gums on the Growth of Probiotic Microorganisms

Bernice D. Karlton-Senaye and Salam A. Ibrahim

Abstract Probiotics are increasingly being used as dietary supplements in functional food products. The ingredients in food contribute to maintaining the growth of probiotics in food products. Gums are polysaccharides used as stabilizers and emulsifiers in foods and could also be used to enhance the growth of probiotics. Thus, the objective of this study was to determine the impact of different gums on the growth of *Lactobacillus reuteri* in laboratory media. Modified M17 media was prepared with 0.5 % (w/v) of one of the following gums: locust bean, carrageenan-locust bean-guar, guar, guar-pectin, carrageenan-maltodextrin, pectin, pectin-carrageenan, and inulin pectin-dextrose. Two liters of modified M17 medium was prepared and divided into ten portions. Batches of 200 ml samples with 1 g of each gum were sterilized at 110 °C for 10 min. Sterilized samples were allowed to cool to 42 °C and then inoculated with *L. reuteri* strains at a final inoculum level of 3 log CFU/mL. The inoculated samples were incubated at 37 °C for 16 h, serially diluted, and plated on lactobacilli MRS agar to obtain final bacterial counts. The results showed higher bacterial counts in samples with gums compared to the control. The bacterial population in the control sample increased from initial counts of 2.78 to 7.16 log CFU/mL whereas samples with pectin increased from 3.3 to 9.0 log CFU/mL. The bacterial population in samples with carrageenan also increased from 3.1 to 8.9 log CFU/mL. The average pH samples at the end of incubation ranged from 5.62 (for sample with pectin) to 6.47 (for control without gum). These findings could lead to the use of pectin and carrageenan-maltodextrin as microbial growth enhancers that could subsequently improve the therapeutic quality of functional food.

B.D. Karlton-Senaye • S.A. Ibrahim (✉)
North Carolina Agricultural and Technical State University, Greensboro, NC 27411-1064, USA
e-mail: ibrah001@ncat.edu

Introduction

Gums are complex polysaccharides extracted from sources such as the endosperm of plant seeds, plant exudates, sea weeds, bacterial and animal sources (Amid and Mirhosseini 2012; Phillips and Williams 2000). Gums are also polymers with hydrophilic ability due to the presence of a hydroxyl bond. The composition and structure of gums enable them to imbibe large amounts of water, forming a gel. It is this hydrophilic characteristic that makes gums useful in the food industry. They are used as stabilizers to improve viscosity and texture by preventing "wheying off" (Schmidt 1994). Gums also contribute fiber, enhance sensory qualities of foods (Riedo et al. 2010), and promote growth in probiotics (Karlton-Senaye et al. 2013). In addition, gums are used in other industries, namely as thickeners in the pharmaceutical, cosmetic, paint, inks, paper, color, and adhesive industries (Riedo et al. 2010).

A prebiotic is defined as a "non-digestible food ingredient that beneficially affects the host by selectively stimulating the growth and/or activity of one or a limited number of bacteria in the colon, and thus improves host health" (Gibson et al. 2004). Techniques that combine both probiotics and prebiotics are called synbiotics. Symbiosis is a combination of probiotics and prebiotics that beneficially affect the host by improving the survival and implantation of live microbial dietary supplements in the gastrointestinal tract, by selectively stimulating the growth of probiotics and/or by activating the metabolism of one or a limited number of health-promoting bacteria, and thus improving host welfare (Gibson and Roberfroid 1995; Iacono et al. 2011). Many natural ingredients have been approved as prebiotics that promote the growth and viability of lactic acid bacteria (LAB) in dairy products (Ranadheera et al. 2010). However, inulin and fructooligosaccharide (FOS) have been the most extensively studied prebiotics (Akalin et al. 2004; Donkor et al. 2007; Gibson et al. 2004; Oliveira et al. 2009).

Probiotic microorganisms promote a healthy gastrointestinal tract. These health benefits include reinforcement of gut mucosal immunity, decreased risk associated with mutagenicity and carcinogenicity, alleviation of lactose intolerance, acceleration of intestinal mobility, hypocholesterolemic effects, reduced duration of diarrhea, prevention of inflammatory bowel disease, prevention of colon cancer, inhibition of intestinal pathogens, and treatment and prevention of allergies (Passos and Claudio 2009). However, it must be noted that these microorganisms are vulnerable to various stressful physical and physiological conditions to which they are subjected during fermentation, storage, and digestion processes. The survival of probiotic bacteria is also dependent on the composition and by-products of the fermentation of food and the interaction between different strains of bacteria (Yeo and Liong 2010). To create the required microbial balance and promote the health functionality of probiotic bacteria, there is a need to consume and maintain a minimum amount of 10^6 CFU/mL of probiotic cultures in food daily (Eamonn 2010). To attain this level, the consumption of 100 g of products containing 10^6–10^7 CFU/mL live cells has been suggested (Yeo and Liong 2010). Gums, like other polysaccharides, contain the necessary food ingredients of carbohydrates, sugars, salts, and minerals that could support the survival of probiotic organisms in food and subsequently in the gut. This level of probiotics creates gastrointestinal balance, positively impacts intestinal health and provides other health benefits. Limited studies have shown the impact of gums on probiotics. For example, one study conducted by Ghasempour et al. (2012) showed that zedo gum did not improve viability of probiotic bacteria although minimum therapeutic levels of 10^6–10^7 CFU/mL were maintained. In several other studies, inulin, being a recognized probiotic, was reported to enhance the growth and viability of probiotic bacteria (Donkor et al. 2007; Ranadheera et al. 2010). Though gums are known to improve the texture of food, limited studies have been done using gums to enhance the growth and viability of probiotics bacteria. The aim of this work was to study the impact of gums on the growth of *Lactobacillus* strains in a laboratory medium.

Materials and Methods

Maintenance of culture. Pure cultures of *L. reuteri* (DSM20016 and SD2112) strains were obtained from the culture collection of the Food Microbiology and Biotechnology Laboratory at North Carolina A&T State University (Greensboro, NC, USA). Nine gums (locust bean, carrageenan-locust bean-guar, guar, guar-pectin, carrageenan-maltodextrin, pectin, pectin-carrageenan, and inulin pectin-dextrose) were kindly provided by TIC Gums (Maryland, GA, USA). Each strain was stored at −80 °C. Sterile 10 mL aliquots of de Man Rogosa and Sharpe (MRS) broth (Sigma Chemical Co.) were inoculated with 1 % (v/v) of each strain and incubated at 37 °C for 16 h. The activated bacterial stains were then used to prepare a mixed culture after three successive transfers.

Preparation of media and treatments. Modified M17 media (Ibrahim et al. 2009) was prepared by dissolving M17 (37.25 g), Yeast extract (5.0 g), beef extract (5.0 g), tryptic soy (5.0 g), and L-cystein (0.1 g) in 1 L of distilled water and divided into ten batches 100 mL each. Gums were added at 0.5 % (w/v) and allowed to dissolve. Sample without gum was the control. The batches of modified M17 media containing each gum were sterilized at 110 °C for 10 min and allowed to cool to 42 °C. After serial dilution, 1 mL of aliquot of inoculum was inoculated into each batch of M17 media containing gum. Samples were surface plated on *Lactobacilli* MRS agar (Sigma Chemical Co.) and incubated at 37 °C for 48 h to determine initial bacterial counts. Batches of 100 mL samples were then incubated at 37 °C for 16 h, plated on *Lactobacilli* MRS agar, and incubated at 37 °C. Plates with 30–300 colonies were counted to determine the final bacterial population. The pH values of the samples were also determined.

Determination of pH values. Each sample was withdrawn after incubation to measure pH. The pH meter (Model 410A, Orion, Boston, MA) was calibrated with pH standard buffers 4.0 and 7.0. After calibration, pH of samples was taken and recorded. Electrodes were rinsed with distilled water after each sample.

Results and Discussion

Bacterial enumeration. Figure 1 shows the impact of nine different gums on the population of *Lactobacillus reuteri* strains. The bacterial population in the control sample increased from initial counts of 2.78 to 7.16 log CFU/mL, whereas samples with pectin increased from 3.3 to 9 log CFU/mL. The bacterial population in samples with carrageenan-maltodextrin also increased from 3.1 to 8.9 log CFU/mL. Yeo and Liong (2010) observed similar trend in their study. Inulin was found to be the least effective in stimulating the growth of *L. reuteri* strains. Donkor et al. (2007) reported a significant initial growth in yogurt samples containing inulin. This could be attributed to the specific strains of *Lactobacillus* and to yogurt itself as a medium of growth. Pectin stimulated the highest bacterial growth among the nine gums tested. Pectin has been reported to be metabolized by many species of human gut flora (Olano-Martin et al. 2002), suggesting that pectin could also be used as a prebiotic.

Changes in pH values. Changes in the pH of MRS containing different gums incubated at 37 °C for 16 h are illustrated in Table 1. The initial and final pH of samples ranged from 5.80 to 6.68, and 5.62 to 6.47, respectively. This range of pH is very close to the manufacturer specifications for each respective gum. The pH remained relatively constant in both the control and samples containing gums during fermentation, with pectin exhibiting the lowest pH (Olano-Martin et al. 2002). This slight decrease in pH could be attributed to the low metabolic activity of the bacteria due to fermentation by *L. reuteri*.

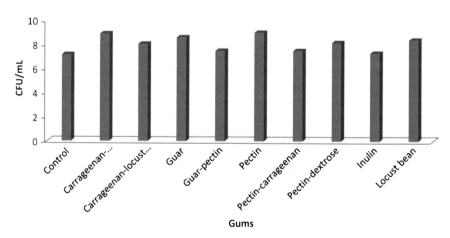

Fig 1 Average CFU/mL of *Lactobacillus reuteri* species in the presence of different gums during incubation at 37 °C for 16 h

Table 1 Average pH values of M17 media containing different gums during 16 h incubation at 37 °C

Gums	pH values	
	Initial	Final
Control	6.68	6.47
Pectin-dextrose	5.96	5.71
Pectin	5.80	5.62
Carrageenan	6.67	6.14
Carrageenan-maltodextin	6.66	6.36
Locust bean	6.62	6.44
Guar	6.51	6.18
Pectin-carrageenan	6.45	6.21
Inulin	6.52	6.26
Carrageenan-locus bean-guar	6.58	6.40

Conclusion

We investigated the impact of different gums on the growth of *Lactobacillus reuteri* strains. The results of our study indicated that the growth of *Lactobacillus reuteri* strains increased in the presence of gums. The presence of gums did not influence the pH to a great extent. Overall, pectin and carrageenan-maltodextrin were found to be the best growth enhancers. However, with further research, pectin and carrageenan-maltodextrin could be used as prebiotics to promote the quality of functional food by enhancing the survival of probiotics in functional foods.

Acknowledgements This work was made possible by grant number NC.X-267-5-12-170-1 from the National Institute of Food and Agriculture and its contents are solely the responsibility of the authors and do not necessarily represent the official view of the National Institute of Food and Agriculture. The gums were supplied by TIC Gums. The authors would like to thank Dr. K. Schimmel for his support while conducting this work.

References

Akalin, A. S., Fenderya, S., & Akbulut, N. (2004). Viability and activity of bifidobacteria in yoghurt containing fructooligosaccharide during refrigerated storage. *International Journal of Food Science & Technology, 39*(6), 613–621.

Amid, B. T., & Mirhosseini, H. (2012). Optimisation of aqueous extraction of gum from durian (Durio zibethinus) seed: A potential, low cost source of hydrocolloid. *Food Chemistry, 132*(3), 1258–1268.

Donkor, O. N., Nilmini, S. L. I., Stolic, P., Vasiljevic, T., & Shah, N. P. (2007). Survival and activity of selected probiotic organisms in set-type yoghurt during cold storage. *International Dairy Journal, 17*(6), 657–665.

Eamonn, M. M. Q. (2010). Prebiotics and probiotics; modifying and mining the microbiota. *Pharmacological Research, 61*(3), 213–218.

Ghasempour, Z., Alizadeh, M., & Bari, M. R. (2012). Optimisation of probitic yoghurt production containing Zedo gum. *International Journal of Dairy Technology, 65*(1), 118–125.

Gibson, G. R., Probert, H. M., Loo, J. V., Rastall, R. A., & Roberfroid, M. B. (2004). Dietary modulation of the human colonic microbiota: Updating the concept of prebiotics. *Nutrition Research Reviews, 17*(2), 259–275.

Gibson, G. R., & Roberfroid, M. B. (1995). Dietary modulation of the human colonic microbiota: Introducing the concept of prebiotics. *Journal of Nutrition, 125*(6), 1401–1412.

Iacono, A., Raso, G. M., Canani, R. B., Calignano, A., & Meli, R. (2011). Probiotics as an emerging therapeutic strategy to treat NAFLD: Focus on molecular and biochemical mechanisms. *The Journal of Nutritional Biochemistry, 22*(8), 699–711.

Ibrahim, S. A., Ahmed, S. A., & Song, D. (2009). Use of Tween 80 to enhance bile tolerance of *Lactobacillus reuteri*. *Milchwissenschaft, 64*(1), 29–31.

Karlton-Senaye, B. D., & Ibrahim, S. A. (2013, July/August). Impact of gums on the growth of probiotics. *Agro Food Industries: Functional food, Nutraceutical, 24*(4), 10–14.

Olano-Martin, E., Gibson, G. R., & Rastall, R. A. (2002). Comparison of the in vitro bifidogenic properties of pectins and pectic-oligosaccharides. *Journal of Applied Microbiology, 93*(3), 505–511.

Oliveira, R. P., Perego, P., Converti, A., & Oliveira, M. N. (2009). The effect of inulin as a prebiotic on the production of probiotic fibre enriched fermented milk. *International Journal of Dairy Technology, 62*(2), 195–203.

Passos, M. L., & Claudio, P. R. (Eds.). (2009). *Innovation in food engineering: New techniques and products* (pp. 601–633). Boca Raton, FL: CRC Press.

Phillips, G. O., & Williams, P. A. (2000). *Handbook of hydrocolloids*. Cambridge, England: Woodhead.

Ranadheera, R. D. C. S., Baines, S. K., & Adams, M. C. (2010). Importance of food in probiotic efficacy. *Food Research International, 43*(1), 1–7.

Riedo, C., Scalarone, D., & Chiantore, O. (2010). Advances in identification of plant gums in cultural heritage by thermally assisted hydrolysis and methylation. *Analytical and Bioanalytical Chemistry, 396*(4), 1559–1569.

Schmidt, K. (1994). Effect of milk proteins and stabilizer on ice milk quality. *Journal of Food Quality, 17*(1), 9–19.

Yeo, S.-K., & Liong, M.-T. (2010). Effect of prebiotics on viability and growth characteristics of probiotics in soymilk. *Journal of the Science of Food and Agriculture, 90*(2), 267–275.

Interaction Between *Bifidobacterium* and Medical Drugs

Temitayo O. Obanla, Saeed A. Hayek, Rabin Gyawali, and Salam A. Ibrahim

Abstract Probiotics are live microorganisms which when administered in adequate amounts confer beneficial health benefits to the host. However, commonly consumed medical drugs may interact with probiotic bacteria and influence their viability and functionality. The objective of this study was to determine the impact of commonly administered medical drugs on the survival of *Bifidobacterium*. Five strains of *Bifidobacterium* (*B. breve*, *B. longum*, *B. infantis*, *B. adolescentis*, and *B. bifidum*) were individually grown in MRS broth at 37 °C for next day use. One tablet of commonly used medical drug (Aleve, Aspirin, Tylenol, Hydro, Lisinopril, Metformin, Metoprolol, or Glipizide) was completely dissolved into batches of 9 mL MRS broth then samples were inoculated with 1 mL of overnight grown cultures. Samples were incubated at 37 °C for 2 h and bacterial populations were determined immediately after exposure to medical drugs and after 2 h of incubation. Our result showed a decrease in bifidobacteria population by an average of 3.0 ± 0.25 log CFU/mL in the presence of tested drugs. Arthritis drug (Aleve, Aspirin, and Tylenol) showed higher killing effect on bifidobacteria compared to other tested drugs. These findings suggested that intake of common medical drugs may decrease the viability of probiotic bacteria and may reflect their contribution to human health. The intake of probiotic dietary supplements and functional foods may reduce the negative effect of medical drugs on probiotics.

Introduction

Gut microflora play important roles in human health and disease prevention. Health benefits of gut microflora include metabolizing polysaccharides, activation of the immune system, reducing the prevalence of atopic eczema later in life, contributing to the inactivation of pathogens in the gut, rheumatoid arthritis, improving the

---

T.O. Obanla • S.A. Hayek • R. Gyawali • S.A. Ibrahim (✉)
North Carolina Agricultural and Technical State University, Greensboro, NC 27411-1064, USA
e-mail: ibrah001@ncat.edu

immune response in elderly people, and regulating the host-signaling pathway (Biagi et al. 2013; Brestoff and Artis 2013; Hayek et al. 2013; Song et al. 2012). *Probiotics could contribute to human's health by improving immune system and improve functionality of gastrointestinal tract against pathogenic organisms without any adverse effects on the host* (Gill et al. 2000; Gill 2003; Song et al. 2012). *Probiotics have direct impact on digestion, which can also increase the nutritional value of fermented dairy products* (Ibrahim et al. 2010; Hayek et al. 2013). However, aging process may affect human gut microflora composition and interact with probiotic functionality (Biagi et al. 2013; Woodmansey et al. 2004). The mechanisms behind the changes in gut microflora as age increases can only be speculated. For example, elderly people had reduced levels of bacteroides and bifidobacteria compared to younger ages (Woodmansey et al. 2004), and bifidobacteria exhibited reduced adhesion to the intestinal mucus of elderly people (Ouwehand et al. 2002). Knowledge of age-related changes in the gastrointestinal tract along with changes in gut microflora are important in the treatment and prophylaxis of diseases, and in maintenance of health among elderly (Woodmansey 2007). Elderly in general have reduced immune functions which could be related to the change in the composition of intestinal microflora (Gill et al. 2001; Woodmansey et al. 2004). *Changes in the gut microflora are associated with immunosenescence and infectious diseases in elderly people* (Brestoff and Artis 2013). *In addition to aging, other factors such as diet, lifestyle, and temporary illnesses could change the gut microflora.*

Due to the increase in the global life expectancy, the health status of elderly people is gaining more attention (Biagi et al. 2013). Estimates on the series of diseases among the elderly population as aging progresses confirmed that up to 85 % of elderly population is likely to be placed on one or more medications at any giving time (Biagi et al. 2013). Even though drug therapy could be essential when caring for patients, drug interaction with other drugs or food or drug side effects must be considered. Drugs may interact with the gut microflora and modify their composition (Gill et al. 2001; Woodmansey et al. 2004). Such that antibiotic drugs which are used to kill pathogenic bacteria will also kill other beneficial bacteria. For example, *Bifidobacterium* strains were found sensitive to several antibiotics including penicillins: penicillin G, amoxicillin, piperacillin, ticarcillin, imipenem, and other anti-Gram-positive antibiotics (Moubareck et al. 2005). Thus, it has been demonstrated that probiotics should be resistant to certain antibiotics in order to survive in the gastrointestinal tract (Moubareck et al. 2005; Woodmansey et al. 2004). However, other medications may also interact with probiotics; survival or functionality, and reduce the health benefits. Elderly patients could be at high risk of having drug interactions with probiotic bacteria since most of them are placed on medications. Arthritis, hypertension, and diabetic medications are among the most commonly used drugs by elderly patients. However, the interactions between these medical drugs and probiotic bacteria have not been investigated. The objective of this study was to determine the effect of commonly administered medications on the survival of bifidobacteria.

Table 1 List of Bifidobacterium strains used in this study

Strain	Original source
B. adolescentis	ATCC 15704
B. bifidum	ATCC 29521
B. longum	ATCC 15708
B. breve	ATCC 15703
B. longum	ATCC 15707

Materials and Methods

Culture preparation and activation. Bifidobacterium stains (Table 1) were obtained from the stock culture collection of Food Microbiology and Biotechnology Laboratory at North Carolina A&T State University, Greensboro, NC. To activate the strains, 100 µL of stock culture was transferred to 5 mL MRS broth and incubated at 37 °C for 24 h. Strains were streaked onto MRS agar and incubated for 48 h at 37 °C then stored at 4 °C. Prior to each experimental replicate, one colony was transferred to 5 mL MRS broth and incubated at 37 °C for 24 h.

Bacterial enumeration. Bacterial populations were determined by plating appropriate dilutions onto MRS agar. In this procedure, samples were individually diluted into serials of 9 mL 0.1 % peptone water solution then 100 µL were surface-plated onto triplicates of MRS agar then incubated at 37 °C for 48 h. Plates having 25–250 colonies were considered for colony counting and bacterial populations were expressed as log CFU/mL.

Drug impact assay. Medical drugs commonly used for arthritis (Aleve, Aspirin, and Tylenol), hypertension (Hydro, Lisinopril, and Metoprolol), and diabetes (Metformin, and Glipizide) were used in this study. Medical drugs were obtained from a local pharmacy at Greensboro, NC, and stored at room temperature. One tablet of each drug was totally crushed then dissolved in 9 mL of MRS broth. Batches of 9 mL MRS without medications served as control. Samples were inoculated with 1 mL of overnight grown strains (containing 10–11 log CFU/mL) and bacterial populations were determined after drug exposure (0 h) and after 2 h of incubation at 37 °C.

Result and Discussion

The interaction between commonly administered medical drugs (Hydro, Glipizide, Lisinopril, Metformin, Metoprolol, Tylenol, Aspirin, and Aleve) and Bifidobacterium strains was studied. Tables 2 and 3 show the survival of Bifidobacterium strains immediately after exposure to medical drugs (0 h) and after 2 h of incubation at 37 °C, respectively. In control samples, Bifidobacterium strains remained at original population with an average of 10.33 ± 0.06 log CFU/mL. The exposure of Bifidobacterium strains to medical

Table 2 Populations of *Bifidobacterium* strains immediately after exposure to medical drugs (0 h)

Strain	Bacterial population (log CFU/mL)								
	Hydro	Glipizide	Lisinopril	Metformin	Metoprolol	Tylenol	Aspirin	Aleve	Control
B. adolescentis	7.62	7.51	7.60	7.60	7.58	6.55	6.53	6.44	10.26
B. bifidum	7.80	7.60	7.70	7.66	7.46	6.46	6.55	6.44	10.31
B. longum	7.95	7.98	8.06	7.85	7.70	6.51	6.58	6.46	10.38
B. breve	7.99	7.97	7.70	7.86	7.60	6.53	6.50	6.47	10.30
B. longum	7.94	7.85	7.80	7.89	7.56	6.60	6.57	6.41	10.40

Table 3 Populations of *Bifidobacterium* strains in the presence of medical drugs after 2 h of incubation at 37 °C

Strains	Bacterial population log CFU/mL							
	Hydro	Glipizide	Lisinopril	Metformin	Metoprolol	Tylenol	Aspirin	Aleve	Control
B. adolescentis	7.58	7.51	7.52	7.54	7.55	6.43	6.39	6.36	10.28
B. bifidum	7.76	7.54	7.65	7.66	7.36	6.40	6.30	6.32	10.34
B. longum	7.85	7.79	7.88	7.71	7.59	6.42	6.22	6.31	10.41
B. breve	7.86	7.84	7.57	7.71	7.53	6.38	6.42	6.23	10.35
B. longum	7.90	7.83	7.76	7.84	7.56	6.43	6.31	6.29	10.43

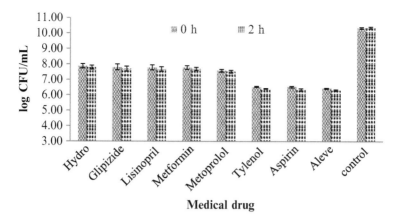

Fig. 1 *Average population* of *Bifidobacterium* strains immediately after exposure to medical drugs (0 h) and after 2 h of incubation at 37 °C. Data points are the average of five strains with standard error bar

drugs has reduced the bacterial population to reach an average of 7.29 ± 0.65 log CFU/mL. There was an average of 3.04 ± 0.59 log CFU/mL reduction in the bacterial populations due to the exposure to medical drugs. Tylenol, Aspirin, and Aleve showed higher killing effect on *Bifidobacterium* strains compared to other tested medical drugs. Populations of *Bifidobacterium* strains were reduced to averages of 6.53 ± 0.05, 6.55 ± 0.02, and 6.44 ± 0.02 due to the exposure to Tylenol, Aspirin, and Aleve, respectively. After 2 h of incubation, the average population of *Bifidobacterium* strains in control was 10.36 ± 0.06 log CFU/mL. In the presence of medical drugs, the average of bacterial population was reduced to reach an average of 7.18 ± 0.69 log CFU/mL. Bacterial populations after 2 h of incubation were similar to that after exposure. Similar effect for each medical drug was obtained on different *Bifidobacterium* strains at 0 and 2 h. These results indicated that common medical drugs could kill *Bifidobacterium* strains after exposure in laboratory medium. However, the effect of medical drugs on different *Bifidobacterium* strains is not strain specific.

Figure 1 shows averages of *Bifidobacterium* strains populations immediately after exposure to medical drugs and after 2 h of incubation at 37 °C. On average, bacterial populations of *Bifidobacterium* strains after 2 h of incubation with medical drugs were similar to that at 0 h. Only slight reduction in the final bacterial population was shown after 2 h of incubation compared to 0 h. Incubation of bifidobacteria with medical drugs was expected to cause additional reduction in the bacterial population. These results may indicate that *Bifidobacterium* strains could die as soon as they are exposed to medical drugs. However, if the strains were able to survive after exposure to medical drugs, no further killing could result due to the presence of medical drugs. On the other hand, the killing effect of medical drugs on *Bifidobacterium* strains may vary among drugs. However, further work is required to determine the effect of different concentrations of medical drugs on the survival of *Bifidobacterium* strains.

The killing effect of medical drugs used for arthritis was higher than that used for hypertension and diabetes medications. The killing effect of arthritis medications could be explained as a side effect of their anti-inflammatory properties. For example, Aspirin is an NSAID (nonsteroidal anti-inflammatory drug), and it is often used to treat arthritis, toothaches, and other pains aggravated by inflammation (Singh 2013). However, Aspirin has been noticed for it is undesirable side effects such as gastrointestinal ulcers, stomach bleeding, and tinnitus, especially in high doses. NSAID drugs could affect the microbial metabolism of gut microbiota and increase the aerobes microbes in elderly (Tiihonen et al. 2008). It is also possible that NSAID drugs may reduce the immune response (Tiihonen et al. 2008). Such anti-inflammatory properties of arthritis drugs may alter the structure of the bacterial cell wall and result in killing of the cell. NSAID drugs are the most prescribed drugs in the world but their use is associated with several side effects including gastrointestinal injury and peptic ulceration (Singh 2013). Aspirin is among the most prominent members of this group of drugs. In addition, most of these drugs are available over the counter in most countries (Singh 2013). In 2001, NSAID drugs accounted for 70 million prescriptions and 30 billion over-the-counter drugs sold in the United States (Singh 2013). In this chapter, we are reporting the killing effect of NSAID drugs on bifidobacteria. These drugs may also interact with other beneficial bacteria in human guts and reduce their viability. Due to the reduction in viable cells of probiotic bacteria in the gut, the health benefits of such beneficial bacteria will be reduced.

Conclusion

The effect of commonly used medical drugs among elderly patients on the survival of *Bifidobacterium* was studied. Our results indicated that common medical drugs could kill *Bifidobacterium* strains in laboratory medium as soon as the bacteria are exposed to the drugs. Further incubation of *Bifidobacterium* in the presence of medical drugs will not reduce the bacteria population. The effect of medical drugs on the survival of *Bifidobacterium* strains may vary among drugs. Drugs used to treat arthritis such as Tylenol, Aspirin, and Aleve have higher killing effect compared to hypertension and diabetes medications. In addition, the effect of medical drugs on *Bifidobacterium* strains is not strain specific. Thus, the result of this research suggested that elderly citizens will need to consume food supplemented with probiotics and take probiotic supplements on a daily basis in order to maintain their health.

Acknowledgment This work was made possible by grant number NC.X-267-5-12-170-1 from the National Institute of Food and Agriculture and its contents are solely the responsibility of the authors and do not necessarily respond to the official view of the National Institute of Food and Agriculture.

References

Biagi, E., Candela, M., Turroni, S., Garagnani, P., Franceschi, C., & Brigidi, P. (2013). Ageing and gut microbes: Perspectives for health maintenance and longevity. *Pharmacological Research, 69*(1), 11–20.

Brestoff, J. R., & Artis, D. (2013). Commensal bacteria at the interface of host metabolism and the immune system. *Nature Immunology, 14*(7), 676–684.

Gill, H. S. (2003). Probiotics to enhance anti-infective defences in the gastrointestinal tract. *Best Practice & Research Clinical Gastroenterology, 17*(5), 755–773.

Gill, H. S., Rutherfurd, K. J., Prasad, J., & Gopal, P. K. (2000). Enhancement of natural and acquired immunity by *Lactobacillus rhamnosus* (HN001), *Lactobacillus acidophilus* (HN017) and *Bifidobacterium lactis* (HN019). *British Journal of Nutrition, 83*(02), 167–176.

Gill, H. S., Rutherfurd, K. J., & Cross, M. L. (2001). Dietary probiotic supplementation enhances natural killer cell activity in the elderly: An investigation of age-related immunological changes. *Journal of Clinical Immunology, 21*(4), 264–271.

Hayek, S. A., Shahbazi, A., Worku, M., & Ibrahim, S. A. (2013). Enzymatic activity of *Lactobacillus* grown in a sweet potato base medium. *British Microbiology Research Journal, 4*(5), 509–522.

Ibrahim, S. A., Alazzeh, A. Y., Awaisheh, S. S., Song, D., Shahbazi, A., & AbuGhazaleh, A. A. (2010). Enhancement of α-and β-galactosidase activity in *Lactobacillus reuteri* by different metal ions. *Biological Trace Element Research, 136*(1), 106–116.

Moubareck, C., Gavini, F., Vaugien, L., Butel, M. J., & Doucet-Populaire, F. (2005). Antimicrobial susceptibility of bifidobacteria. *Journal of Antimicrobial Chemotherapy, 55*, 38–44.

Ouwehand, A. C., Salminen, S., & Isolauri, E. (2002). Probiotics: An overview of beneficial effects. *Antonie Van Leeuwenhoek, 82*(1–4), 279–289.

Singh, S. (2013). An overview of NSAIDs used in anti-inflammatory and analgesic activity and prevention gastrointestinal damage. *Journal of Drug Discovery and Therapeutics, 1*(8), 41–51.

Song, D., Ibrahim, S., & Hayek, S. (2012). Recent application of probiotics in food and agricultural science. In E. C. Rigobelo (Ed.), *Probiotics* (1st ed., pp. 3–36). New York: InTech.

Tiihonen, K., Tynkkynen, S., Ouwehand, A., Ahlroos, T., & Rautonen, N. (2008). The effect of ageing with and without non-steroidal anti-inflammatory drugs on gastrointestinal microbiology and immunology. *British Journal of Nutrition, 100*(1), 130–137.

Woodmansey, E. J. (2007). Intestinal bacteria and ageing. *Journal of Applied Microbiology, 102*(5), 1178–1186.

Woodmansey, E. J., McMurdo, M. E., Macfarlane, G. T., & Macfarlane, S. (2004). Comparison of compositions and metabolic activities of fecal microbiotas in young adults and in antibiotic-treated and non-antibiotic-treated elderly subjects. *Applied and Environmental Microbiology, 70*(10), 6113–6122.

Functional Food Product Development from Fish Processing By-products Using Isoelectric Solubilization/Precipitation

Reza Tahergorabi and Salam Ibrahim

Abstract Currently, the western diet typically includes high salt content and low omega-3 fatty acids and fiber. Such a diet enhances risk factors for cardiovascular disease. The fish processing industry generates large amounts of by-products, and these by-products have the potential for use as a heart-healthy human food. One way to increase the utilization of these fish processing by-products is by using isoelectric solubilization/precipitation (ISP) to recover myofibrillar proteins and separate them from pigments and fat in order to increase their utilization for the development of functional food products. Unlike mechanical meat recovery such as deboning of fish meat, ISP allows selective pH-induced water solubility of meat proteins with concurrent separation of lipids and the removal of materials not intended for human consumption such as bones, skin, and scale. As a result, fish protein isolate gels made with nutraceutical additives (omega-3 fatty acids-rich oil, dietary fiber, and a salt substitute) provide a means to achieve the desired biochemical effects of these nutrients without the need for dietary supplements and medications or a major change in dietary habits.

Introduction

There has been confusion over the actual definition of aquatic food processing "by-products." Currently, the most common definition of "by-products" is all of the edible or inedible materials left over following processing of the main product (Gehring et al. 2010). A typical example is fish filleting to recover boneless and skinless marketable fillets. The fillets would be considered to be the main product, and the frames, heads, and viscera would be typical "by-products." It would be misleading to name these by-products as "waste" since at the time of filleting the quality of fish meat left on the frame (i.e., fish bones) and in the heads is not compromised (Strom and Eggum 1981). If proper meat recovery technology is successfully applied, the recovered meat can result in added revenue for a processor

R. Tahergorabi (✉) • S. Ibrahim
North Carolina Agricultural and Technical State University, Greensboro, NC 27411, USA
e-mail: rtahergo@ncat.edu

© Springer International Publishing Switzerland 2016
G.A. Uzochukwu et al. (eds.), *Proceedings of the 2013 National Conference on Advances in Environmental Science and Technology*,
DOI 10.1007/978-3-319-19923-8_18

as well as reduce environmental stress associated with the disposal of the processing by-products.

Unlike mechanical meat recovery such as deboning of fish meat processing by-products, isoelectric solubilization/precipitation (ISP) allows selective, pH-induced water solubility of meat proteins with concurrent separation of lipids and the removal of materials not intended for human consumption such as bones and skin (Tahergorabi et al. 2011).

Functional food product development. Muscle food products are essential in the diet of developed countries. The principal components of muscle food products, besides water, are proteins and fats, with a substantial contribution of vitamins and minerals with a high degree of bioavailability. Both fish and its associated products can be modified by adding ingredients considered beneficial for health or by eliminating or reducing components that are considered harmful. In this way, a series of foods can be obtained which, without altering their base, are considered to be healthy.

People who have adopted such diets have benefitted due to a much lower risk of heart disease (McCullough et al. 2000). However, such a prudent diet is not typical of what consumers in Western countries eat (McCullough et al. 2000; Kennedy et al. 1995). It appears that consumers today are less likely to invest in long-term health if taste and convenience are compromised. Food industries are aware of this, and market some of their foods with health claims. Indeed, it has been shown that health claims on foods have a positive influence on consumers' perception of the healthiness of foods. Thus, functional foods in the form of palatable and ready-to-use foods that suggest short- or long-term health benefits have a huge market and health potential.

The western diet is characterized by the increased dietary intake of saturated fat, ω-6 fatty acids (ω-6 FAs), and trans-FAs, as well as a decreased intake of ω-3 FAs. As a result, the ratio of ω-6 FAs to ω-3 FAs is at 15–20 to 1, instead of the suggested 1 to 1 (Eaton et al. 1988; Simopoulos 1999). According to the American Heart Association, cardiovascular disease (CVD) has had an unquestioned status as the number one cause of death in the United States since 1921 (American Heart Association 2009). In 2004, the Food and Drug Administration (FDA) approved a health claim for reduced risk of CVD for foods containing ω-3 PUFAs, mainly EPA and DHA (FDA 2004). This change provided marketing leverage for functional foods fortified with ω-3 PUFAs and initiated the development of food products addressing the diet-driven CVD. Since seafood products developed from the ISP-recovered fish protein isolate would be formulated products associated with aquatic sources, they are a logical vehicle for increasing the consumption of ω-3 PUFAs. Such enhanced heart-healthy products would also address the diet-driven CVD without the need for dietary supplements in a pill or capsule form.

Tahergorabi et al. (2013) used the ISP processing technique to recover protein isolates from whole gutted rainbow trout (bone-in, skin- and scale-on) as a model for fish processing by-products. Fortification of the ISP-recovered fish protein isolate with ω-3 polyunsaturated fatty acids (PUFAs)-rich oils (flaxseed, fish, algae, krill, and blend) resulted in an increased ($p < 0.05$) content of alpha-linolenic

(ALA, 18:3ω-3), eicosapentaenoic (EPA, 20:5ω-3), and docosahexaenoic acids (DHA, 22:6ω-3) in the cooked protein isolate gels. The extent of the PUFAs increase, ω-6/ω-3 FAs, and unsaturated/saturated FAs ratios, as well as the indices of thrombogenicity and atherogenicity depended on specific ω-3 PUFAs-rich oils used to fortify the protein isolate gels. Lipid oxidation and protein degradation were slightly higher ($p < 0.05$) in the ω-3 PUFAs fortified gels than in the control gels. However, all gels were within acceptable range. This study thus indicates a potential application for protein isolates recovered with ISP from fish processing by-products or whole fish without prior filleting. Fortification of the ISP protein isolate with ω-3 PUFAs-rich oils allows the development of a functional seafood product.

There is evidence that excessive dietary salt intake raises blood pressure, which is the major cause of CVD (Intersalt Cooperative Research Group 1988; Lifton 1996; World Health Organization 2002). The current average dietary sodium intake in the United States exceeds 3400 mg/day, which is considerably higher than the recommended maximum intake level of 2300 mg/day (United States Department of Agriculture 2011). The Centers for Disease Control and Prevention (CDC) have proposed a limit of 1500 mg/day (CDC 2009).

Normally, NaOH and HCl are used to dissolve and precipitate, respectively, fish muscle proteins in isoelectric solubilization/precipitation (ISP), thereby contributing to increased Na content in the recovered fish protein isolates (FPI). Substitution of NaOH with KOH may decrease the Na content in FPI and, thus, allow the development of reduced-Na seafood products. In this regard, Tahergorabi et al. (2012a) recovered FPI with ISP using NaOH or KOH. In order to develop a nutraceutical seafood product, the FPI was extracted with a NaCl or KCl-based salt substitute and subjected to cold- or heat-gelation. In addition, standard nutraceutical additives (ω-3 fatty acids-rich oil and dietary fiber) along with titanium dioxide (TiO_2) were added to FPI. Color, texture, dynamic rheology, Na and K content, and lipid oxidation of the FPI gels were compared to commercial Alaska pollock surimi gels. FPI gels had greater ($p < 0.05$) whiteness, good color properties (L*a*b*), and generally better textural properties when compared to surimi gels. A reduction ($p < 0.05$) of Na content and simultaneous increase ($p < 0.05$) in K content of FPI gels was achieved by the substitution of NaOH with KOH during ISP and NaCl with the KCl-based salt substitute during formulation of the FPI paste. These results indicate that KOH can replace NaOH to recover FPI from whole gutted fish for subsequent development of nutraceutical seafood products tailored for reduction of diet-driven cardiovascular disease.

In contrast to sodium, dietary fiber has cardiovascular benefits (Anderson and Ma 2009). The American diet is deficient in fiber with an average intake of only 15 g/day (Dietary Guidelines Advisory Committee 2010). The Institute of Medicine recommends fiber intake to be 25–38 g/day. Dietary fiber has been defined as remnants of plant edible parts and analogous carbohydrates that are resistant to digestion and absorption in humans. It includes polysaccharides, oligosaccharides, lignin, and associated plant substances that benefit human health (Bodner and Sieg 2009). Finally, a three-prong strategy (1—fiber, 2—ω-3 PUFAs, and 3—salt substitute) has been proposed to address the diet-driven CVD (Tahergorabi et al. 2012a, b). This would render a food product that aligns with the dietary recommendations posed by the American Heart Association.

Conclusions

The importance of functional foods, nutraceuticals, and other natural health products has been well recognized in connection with health promotion, disease risk reduction, and reduction in health care costs. Isoelectric solubilization/precipitation (ISP) can be used to recover a fish protein isolate (FPI) from fish processing by-products. FPI gels made with nutraceutical additives (ω-3 fatty acids-rich oil, dietary fiber, and salt substitute) thus provide a means to achieve desired biochemical effects of these nutrients without the need for ingestion of dietary supplements, medications, or a major change in dietary habits.

References

American Heart Association. (2009). *Heart disease and stroke statistics—2009 update*. Retrieved June 12, 2012, from http://www.americanheart.org/presenter.jhtml?identifier=3037327.

Anderson, B. M., & Ma, D. W. (2009). Are all n-3 polyunsaturated fatty acids created equal? *Lipids in Health and Disease, 8*, 33.

Bodner, J. M., & Sieg, J. (2009). Fiber. In R. Tarté (Ed.), *Ingredients in meat products: Properties, functionality and applications* (pp. 83–108). New York: Springer.

Centers for Disease Control and Prevention. (2009). *Americans consume too much salt*. Retrieved March 07, 2011, from http://www.cdc.gov/media/pressrel/2009/r090326.htm.

Dietary Guidelines Advisory Committee. (2010). *Report of the dietary guidelines advisory committee on the dietary guidelines for Americans, 2010, to the secretary of Agriculture and the Secretary of Health and Human Services*. Washington, DC: GPO. Retrieved August 15, 2012, from http://www.cnpp.usda.gov/Publications/DietaryGuidelines/2010/PolicyDoc/PolicyDoc.pdf.

Eaton, S. B., Konner, M., & Shostak, M. (1988). Stone agers in the fast lane: Chronic degenerative diseases in evolutionary perspective. *American Journal of Medicine, 84*, 739–749.

Food and Drug Administration. (2004). Retrieved June 14, 2012, from www.fda.gov/SiteIndex/ucm108341.htm.

Gehring, C. K., Gigliotti, J. C., Moritz, J. S., Tou, J. C., & Jaczynski, J. (2010). Functional and nutritional characteristics of proteins and lipids recovered by isoelectric processing of fish byproducts and low-value fish-a review. *Food Chemistry, 124*(2), 422–431.

Intersalt Cooperative Research Group. (1988). Intersalt: An international study of electrolyte excretion and blood pressure. Results for 24 hour urinary sodium and potassium excretion. *British Medical Journal, 297*, 319–328.

Kennedy, E. T., Ohls, J., Carlson, S., & Fleming, K. (1995). The healthy eating index: Design and applications. *Journal of American Dietetic Association, 95*(10), 1103–1108.

Lifton, R. P. (1996). Molecular genetics of human blood pressure variation. *Science, 272*, 676–680.

McCullough, M. L., Feskanich, D., Stampfer, M. J., et al. (2000). Adherence to the Dietary Guidelines for Americans and risk of major chronic disease in women. *American Journal of Clinical Nutrition, 72*, 1214–1222.

Simopoulos, A. P. (1999). New products from the agri-food industry: The return of n-3 fatty acids into the food supply. *Lipids, 34*(1), S297–S301.

Strom, T., & Eggum, B. O. (1981). Nutritional value of fish viscera silage. *Journal of the Science of Food Agriculture, 32*, 115–120.

Tahergorabi, R., Beamer, S., Matak, K. E., & Jaczynski, J. (2011). Effect of isoelectric solubilisation/precipitation and titanium dioxide on whitening and texture of proteins recovered from dark chicken-meat processing by-products. *LWT–Food Science and Technology, 44*(4), 896–903.

Tahergorabi, R., Beamer, S. K., Matak, K. E., & Jaczynski, J. (2012a). Physicochemical properties of surimi gels with salt substitute. *LWT–Food Science and Technology, 48*, 175–181.

Tahergorabi, R., Beamer, S., Matak, K. E., & Jaczynski, J. (2013). Chemical properties of ω-3 fortified gels made of protein isolate recovered with isoelectric solubilisation/precipitation from whole fish. *Food Chemistry, 139*, 777–785.

Tahergorabi, R., Sivanandan, L., & Jaczynski, J. (2012b). Dynamic rheology and endothermic transitions of proteins recovered from chicken-meat processing by-products using isoelectric solubilization/precipitation. *LWT–Food Science and Technology, 46*(1), 148–155.

United States Department of Agriculture. (2011). *Dietary guidelines for Americans, 2010*. Retrieved February 08, 2012, from http://health.gov/dietaryguidelines/2010.asp.

World Health Organization. (2002). *World health report 2002. Reducing risks, promoting healthy life*. Geneva, Switzerland: World Health Organization. Retrieved February 20, 2010, from http://www.who.int/whr/2002.

Part IV
Sustainable Energy

Optimizing the Design of Chilled Water Plants in Large Commercial Buildings

Dante' Freeland, Christopher Hall, and Nabil Nassif

Abstract Design of chilled water plants has very large impact on building energy use and energy operating cost. The chapter proposes procedures and analysis techniques for energy efficiency design of chilled water plants. The approach that leads to optimal design variables can achieve a significant saving in cooling cost. The optimal variables include piping sizing, chilled water temperature difference, and chilled water supply temperature. The objective function is the total cooling energy cost. The proposed design method depends on detailed cooling load analysis, head and energy calculations, and optimization solver. The pump head calculations including piping, all fittings, valves, and devices are achieved by using the Darcy–Weisbach Equation and given flow parameters. The energy calculations are done by using generic chiller, fan, and pump models. The method is tested on an existing office two-storey building with a packaged air-cooled chiller. A whole building energy simulation model is used to generate the hourly cooling loads, and then the optimal design variables are found to minimize the total energy cost. The testing results show this approach will achieve better results than rules-of-thumb or traditional design procedures. The cooling energy saving could be up to 10 % depending on particular projects.

Introduction

The development of a software that can optimize the energy usage of the chilled water plant is a tool that will have a significant effect on building energy consumptions and costs (Nassif et al. 2005, 2014; Duda 2012; Taylor 2011; Taylor and McGuire 2008; Hydeman and Zhou 2007; Durkin 2005). Through rigorous programming this software has begun to take form and show the great benefits expected from optimizing several of the key components of the chilled water

---

D. Freeland • C. Hall • N. Nassif (✉)
Department of CAAE Engineering, North Carolina A&T State University, Greensboro, NC 27411, USA
e-mail: nnassif@ncat.edu

plant. By simulating the energy load of a known building and applying it to the aforementioned program, an experiment was created to test how effective a chilled water plant optimization program could really become.

Development of Chilled Water Optimization Program. In developing any new program, there must be a goal set out to be achieved and detailed parameters required to meet this target. The goals looked to be achieved from the program were to find the optimal pump, fan, chiller, and total power for a given building. The key parameters that would allow for these outputs were optimal pipe sizing, chilled water temperature difference, and chilled water supply temperature. Having these factors allowed for thermodynamic and fluid mechanics equations to be introduced to the program. This information was combined into a MATLAB program shown in Fig. 1 below.

Building Energy Simulation. The chilled water plant optimization program being created requires data inputs to be placed into the program in parallel to it running and putting out optimal results. These inputs include the hourly results of a year's timespan of the measured outdoor dry/wet bulb temperatures, along with the total

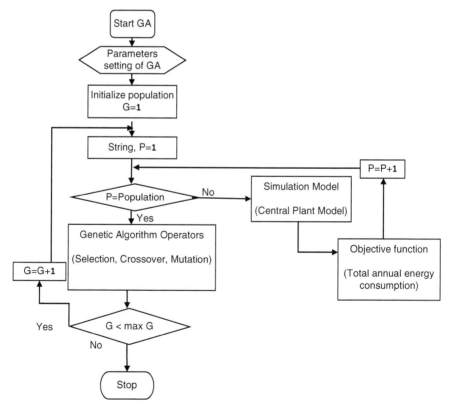

Fig. 1 This is a figure of the chiller plant optimization code in MATLAB

building cfm readings. This data is found by using the DOE software interface called eQuest. When the chosen building is uploaded into the eQuest software, it must be programmed to comply with the climate zone being tested. Next, an energy analysis of the building over a calendar year is created and a report of the data is exported to a Microsoft Excel spreadsheet. The data from this sheet serves as the missing variables from equations in the chilled water plant optimization program. In plugging the excel file into the newly created optimization program, the optimized energy results serve as an output from the program.

Rule of Thumb vs. Optimization. Through centuries of building structures, certain estimates have been established as good points to design new buildings. These estimates have become rule of thumb and have slowed down how buildings can be fully optimized to save as much energy as possible (ASHRAE Standard 90.1 2013). Some of the rule of thumb applications in a chilled water plant focus on: the pipe sizing, chilled water temperature difference, and chilled water supply temperature. Using the typical rule of thumb application for pipe sizing will have the pipe fluctuating between several different sizes based upon the usage. Using this for the chilled water temperature difference and chilled water supply temperature will result in approximately 10 and 45 °F, respectively. These rule of thumb applications are accurate but can be optimized to me more effectively. The optimized pipe size designed for in the program caters to the minimal size needed for the system to function properly based on the optimized result of the chilled water temperature difference. When the system is optimized, a chilled water temperature difference can be expected to be about twice the temperature of the typical rule of thumb application. An optimized chilled water supply temperature will be above the 45 °F rule of thumb, completing the cycle of optimizations.

Methodology

In order to find the optimal design for the chilled water plant a building in Greensboro, NC was taken for analysis. The building's energy consumption was examined using the eQuest energy analysis software. The chilled water plant optimization program code was written in conjunction with this software in order to find the optimal design pipe sizing, chilled water temperature difference, and chilled water supply temperature. The building peak load, total supply cfm, and indoor and outdoor temperature of the specific building region collected from the eQuest simulation were used as input data within the optimization program that was created. By inserting these input values for the different ASHRAE climate zones (ASHRAE Standard 90.1 2013), the optimization program code will determine the optimal designs conditions as well as the pump power, fan power, chiller power, and overall power. These results will be compared to a baseline design, which uses today's nonoptimal design conditions to show the energy savings that can be made.

Data

Optimal vs. Nonoptimal Conditions: Pipe Sizing

See Fig. 2

Chilled Water Temperature Difference

See Fig. 3

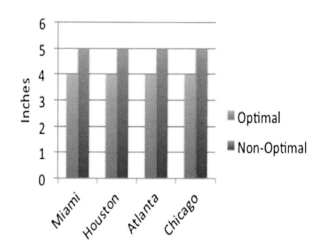

Fig. 2 This graph shows the difference in optimal and nonoptimal design for the pipe size of the building. In the nonoptimal design, a 5-in. pipe size was considered to be the best fit based on sizing charts. When calculated through the chilled water plant optimization program, it was found that only a 4-in. pipe size was actually needed for the system to run efficiently

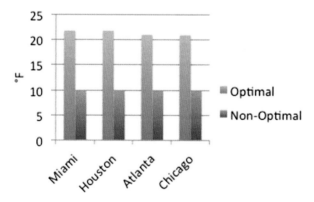

Fig. 3 This graph shows the difference in optimal and nonoptimal design for the chilled water temperature difference of the building. In the nonoptimal design, the temperature difference is 10 °F, while when tested for optimal design through the program the temperature difference was around 22 °F. This increase in chilled water temperature difference allows for a decrease in flow within the system, saving energy and decreasing the pipe size for the system

Chilled Water Supply Temperature

See Fig. 4

Chiller Power

See Fig. 5

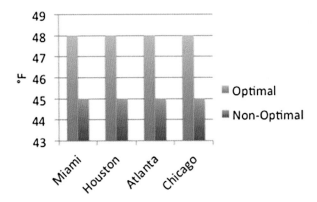

Fig. 4 This graph shows the difference in optimal and nonoptimal design for the chilled water supply temperature. In the nonoptimal design, the supply temperature is 45 °F, while when tested for optimal design through the program the supply temperature was around 48 °F. This increase in chilled water supply temperature will allow for the water to not have to be cooled down to as low a temperature, saving energy

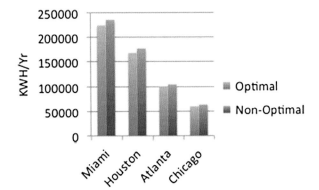

Fig. 5 This graph shows the difference in optimal and nonoptimal design for the chiller power used in the system. The chilled water plant optimization program calculates the amount of chiller power that will be consumed over a year considering the optimal and nonoptimal variables used

Fan Power

See Fig. 6

Pump Power

See Fig. 7

Total Power

See Fig. 8

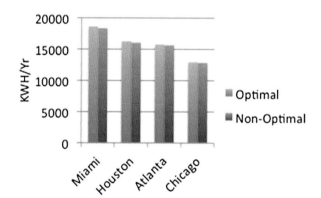

Fig. 6 This graph shows the difference in optimal and nonoptimal design for the fan power used in the system. The chilled water plant optimization program calculates the amount of fan power that will be consumed over a year considering the optimal and nonoptimal variables used

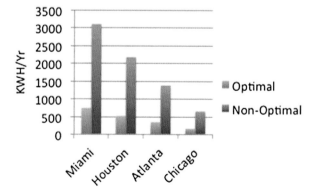

Fig. 7 This graph shows the difference in optimal and nonoptimal design for the pump power used in the system. The chilled water plant optimization program calculates the amount of pump power that will be consumed over a year considering the optimal and nonoptimal variables used

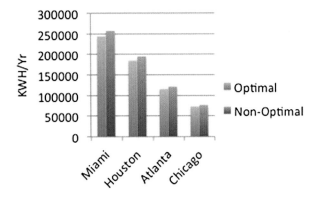

Fig. 8 This graph shows the difference in optimal and nonoptimal design for the total power used in the system. The chilled water plant optimization program calculates the amount of total power that will be consumed over a year considering the optimal and nonoptimal variables used

Data Analysis

When the building energy consumption data from eQuest was run through the chilled water plant optimization program that was created, the results proved a great amount of energy that could be saved by way of optimization. The nonoptimal pipe size of 5 in typically used with a building with this amount of load could be dropped to 4 in. The nonoptimal chilled water temperature difference and chilled water return temperature, 10 °F and 45 °F, respectively, were increased which in turn saves on the energy consumption. With these three parameters changed, a great effect was seen between the optimal and nonoptimal design conditions. In the optimal design conditions, there was a great amount of energy reduction seen in the pump power, chiller power, and total power yearly usage. There however, was an increase in the fan power usage in a year, which can be accredited to the need to add to the rows of cooling coils in the air-handling unit do to the chilled water temperature difference. This increase was minor and the overall power consumption still displayed significant savings.

Conclusion

The optimization of chilled water plants is a subject that needs to be addressed. In this program, three parameters were optimized in order to receive the most potential energy savings in a year. The pipe sizing, chilled water temperature difference, and the chilled water supply temperature all can be considered as major factors that effect the energy consumption of the chilled water plant. The results from the chilled water plant optimization program show that the building in each of the four ASHRAE climate zones run through the program would save around 5 % of the total power consumption in the year. These savings indicate that a substantial amount of money can be saved over the course of the building's life. The impact of these findings shows that other parameters within the chilled water plant should be examined in order to achieve optimal savings. In the future, this building will be tested in the rest of the ASHRAE climate zones to determine the savings in these regions. There will also be a cost saving analysis added within the chilled water plant optimization program in order to predict the annual cost savings for the building in any climate zone. These additions will add to the overall usage of the optimization program and bring it closer to commercial use.

References

ASHRAE Standard 90.1 (2013). *Energy standard for buildings except low-rise residential buildings* (90.1 I-P ed.). Atlanta, GA: ASHRAE.
Duda, S. W. (2012). Easy to use methods for multi-chiller plant energy and cost evaluation. *ASHRAE Journal, 118*, 1–8.
Durkin, T. H. (2005). Evolving design of chilled water plant. *ASHRAE Journal, 47*, 40–50.
Hydeman, M., & Zhou, G. (2007). Optimizing chilled water plant control. *ASHRAE Journal, 49*, 45–54.
Nassif, N., Hall, C., & Freeland, D. (2014). Optimal design and control of ice thermal storage system for a typical chilled water plant. *ASHRAE Transactions, 120*(1).
Nassif, N., Kajl, S., & Sabourin, R. (2005). Optimization of HVAC control system strategy using two-objective genetic algorithm. *HVAC&R Research, 11*(3), 459–486.
Taylor, S. T. (2011). Optimizing design & control of chilled water part 3: Pipe sizing and optimizing ΔT. *ASHRAE Journal, 3*, 22–34.
Taylor, S. T., & McGuire, M. (2008). Sizing pipe using life-cycle costs. *ASHRAE Journal, 1*, 24–32.

Optimizing Ice Thermal Storage to Reduce Energy Cost

Christopher L. Hall, Dante' Freeland, and Nabil Nassif

Abstract Thermal energy storage includes a number of technologies that store thermal energy in energy storage tanks for later use. These applications include the production of ice or chilled water at night which is then used to cool the building during the day. Unfortunately, thermal storage may not provide the expected load shifting or the cost saving if not designed or operated properly. This research discusses the optimal design of ice thermal storage (ITS) and its impact on energy consumption, demand, and total energy cost. The emphasis on the use of ITS as an effort to reduce energy costs lies within transferring the time of most energy consumption from on-peak to off-peak periods. Multiple variables go into the equation of finding the optimal use of ITS, and they are all judged with the final objective of minimizing monthly energy costs. This research discusses the optimal design of ITS and its impact on energy consumption, demand, and total energy cost. A tool for optimal design of ice storage is developed, considering variables such as chiller and ice storage sizes along with ice charge and discharge times. Detailed simulation studies using a real office building located near Orlando, FL including the utility rate structure are presented. The study considers the effect of the ITS on the chiller performance and the associated energy cost and demonstrates the cost saving achieved from optimal ice storage design. A whole building energy simulation model is used to generate the hourly cooling load for both the design day and the entire year. Other collected variables such as condenser entering water temperature, chilled water leaving temperature, outdoor air dry bulb and wet bulb temperatures are used as inputs to a chiller model based on DOE-2 chiller model to determine the associated cooling energy use. The results show a significant cost energy saving can be obtained by optimal ice storage design through using the tool proposed in this chapter.

C.L. Hall • D. Freeland • N. Nassif (✉)
Department of Civil and Architectural Engineering, North Carolina A&T State University, Greensboro, NC, USA
e-mail: nnassif@ncat.edu

Introduction

Thermal energy storage includes a number of technologies that store thermal energy in energy storage tanks for later use. These applications include the production of ice, chilled water, or eutectic solution at night which is then used to cool the building during the day. The ice thermal storage (ITS) is one of thermal energy storage technologies that is widely used in many countries to reduce electrical power or energy costs by moving the cost of cooling buildings from expensive "on-peak" periods to cheaper "off-peak" periods (Sebzali and Rubini 2007; Solberg and Harshaw 2007; Montgomery 1998). The cool-energy is usually stored in the form of ice during the nighttime and used in the daytime. Many studies demonstrate the benefits of ice storage and how the thermal storage can shift the cost of electricity from on-peak to off-peak periods, thus reducing demand and energy charges (Nassif et al. 2013; Yau and Rismanshi 2012; Zhou et al. 2005; MacCracken 2003, 2004; Silvetti 2002; Dincer 2002). Unfortunately, thermal storage may not provide the expected load shifting or the cost saving if not designed or operated properly. The chapter discusses the optimal design of ITS and its impact on energy consumption, demand, and total energy cost. A tool for optimal design of ice storage is developed, considering variables such as chiller and ice storage sizes and charging and discharge times. The tool requires the hourly cooling load that can be obtained from any available energy simulation software. It also requires an optimization algorithm to solve the optimization process. Although there may be many optimization methods that could be used for solving the optimization problem, the genetic algorithm (GA) inspired by natural evolution (Goldberg 1989; Deb 2001) is used. The genetic algorithm is a method for solving both constrained and unconstrained optimization problems that is based on natural selection, the process that drives biological evolution. The GA is successfully applied to a wide range of applications including HVAC system control and design (Nassif 2012; Kusiak et al. 2011; Xu et al. 2009; Mossolly et al. 2009; Nassif et al. 2005).

Detailed simulation studies using real office building located near Orlando, FL including utility rate structure are presented. The study considers the effect of the ITS on the chiller performance and the associated energy cost and demonstrates the cost saving achieved from optimal ice storage design. A whole building energy simulation model eQuest is used to generate the hourly cooling load for both design day and entire year. Other collected variables such as condenser entering water temperature, chilled water leaving temperature, outdoor air dry bulb and wet bulb temperatures are used as inputs to a chiller model based on DOE-2 chiller model to determine the associated cooling energy use. As the main objective of the ITS is to shift the energy use from on-peak period to off-peak period, it is very important to examine the local utility cost structure to identify if the ice storage is cost-effective. A particular utility structure is used as described later in this chapter.

Methodology

To size the chiller, cooling load analysis is generally performed. Traditionally without ice storage, a factor of safety is added to the calculated load. The safety factor could be up to 20 %, and it is acceptable per (ASHRAE Standard 90.1-2004) to oversize the chiller to 115 % of load as a factor of safety (ANSI/ASHRAE/IESNA Standard 90.1-2004). As an alternative, rather than adding the factor of safety, it can be added as ice-storage capacity. In fact, the chiller can size 20–25 % less than the cooling analysis when ice storage is installed. Optimal design of chiller and ice storage is then necessary to achieve the optimal performance. In addition, the optimal discharging and charging times are other important factors to achieve optimal performance. The optimization tool is then developed as shown in Fig. 1 to find the optimal chiller and ice storage sizes and discharging and charging periods. The inputs required are the hourly cooling loads, utility cost structure, and outdoor air conditions. The cooling load and the outdoor air conditions could be generated by any energy simulation software and exported to be used as input. The cost structure should be obtained from the local utility company and should include the cost of kWh and per peak demand kW during the peak and off-peak periods.

The simulation model calculates the hourly cooling power by the chiller model for the charging, discharging, and normal chiller operating periods, and then determines the monthly and whole year energy consumption and associated cost. An optimization algorithm is needed to solve the optimization problem. The problem variables as output of the recommended process are discharging period, charging period, chiller size, and ITS size. The objective function is the annual energy cost.

The optimization seeks to determine the optimal ITS and chiller design to reduce the annual cooling energy cost. The problem variables are (1) chiller size, (2) ITS size, (3) discharging period, and (4) charging period. The objective function is the

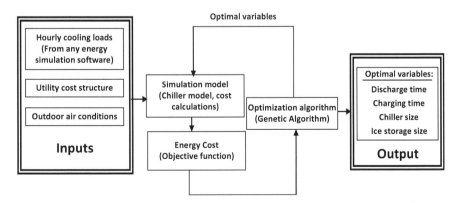

Fig. 1 Recommended optimization tool for optimal design of chiller and ice storage sizes, discharging and charging periods

annual cooling energy cost. The constraints result from restrictions on the size and operation of the central plant. They cover the lower and upper limits of design variables, such as the maximum and minimum size of the chiller, and discharging and charging periods, e.g., discharging period starts only during the peak period, and charging period should start before the occupied period.

The genetic algorithm is used to solve the optimization problem. The genetic algorithm is a method for solving both constrained and unconstrained optimization problems that is based on natural selection, the process that drives biological evolution. The genetic algorithm repeatedly modifies a population of individual solutions. At each step, the genetic algorithm selects individuals at random from the current population to be parents and uses them to produce the children for the next generation. Over successive generations, the population "evolves" toward an optimal solution. The GA starts with a random generation of the initial population (initial solution) and ends with the optimal solutions including the optimal variables. The problem variables represent an individual solution in the population. The performance or objective function of each individual of the first generation is estimated. The second generation is generated using operations on individuals such as selection, crossover, and mutation, in which individuals with higher performance (fitness) have a greater chance to survive. The performance of each new individual is again evaluated. The process is repeated until the maximum number of generations is reached. In this study, the GA algorithm from the optimization tool available in MATLAB® is used (MATLAB 2013).

To test the recommended procedure, an existing 40,000 ft^2 (3716 m^2) office building located near Orlando is selected. The building is occupied from 8:00 AM to 5:00 PM, and the HVAC system turns on at 7:00 AM and turns off at 6:00 PM. The space air is conditioned by typical variable air volume (VAV) systems with chilled water supplied by one screw chiller along with six ice thermal storages. The central plant piping configuration, including pumps, chiller, ice storage, and heat exchanger is illustrated in Fig. 2. The building is served by a 150 t (528 kW) chiller. A total of five pumps circulate either water or a glycol mixture. The piping configuration consists mainly of five water loops (1) primary loop, (2) secondary loop, (3) ITS, (4) heat exchanger (HX), and (5) condenser water loop. In the primary, ITS, and HX loops, there are three pumps circulating glycol water solution through the chiller, ice storage, and heat exchanger, respectively. In the secondary loop, a pump equipped with variable speed drive circulated chilled water to the nine AHUs. In the condenser water loop (not shown in Fig. 2), the pump operates at constant speed to circulate condenser water to the cooling tower when the chiller is operating. There are many operating strategies that could be applied to charging or discharging the storage. The operating strategies include partial or full storage. A full storage strategy is considered for the simulation as it is the one adapted in the existing system. Due to the utility structure considered, the peak period is from 12:00 PM to 5:00 PM; therefore, the ITS is sized to cover this particular period. In this chapter, the following cost structure is assumed. The energy cost if there is no ITS is $0.06 per kWh and $9 per peak demand kW. The cost with ITS is $0.08 per kWh plus $9 per peak demand kW during the peak period and $0.05 per kWh

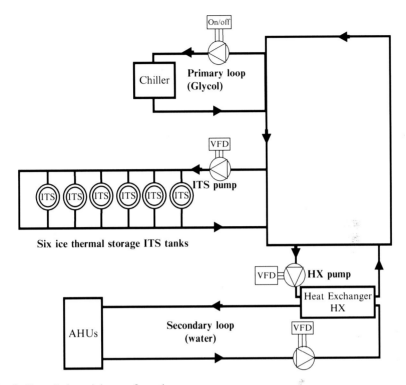

Fig. 2 Central plant piping configuration

during off peak. The peak period is from 12:00 PM to 5:00 PM. It should be noted that different cost structures could be applied in other areas and readers need to check their local utility.

Results

The energy simulation model eQUEST runs for the selected building to determine hourly cooling loads for the design cooling day and for the whole year. Using those cooling loads, the optimization process as shown in Fig. 1 then runs to determine the optimal variables including the chiller and ice storage sizes and charging and discharge times.

The simulations without ITS and with near optimal and nonoptimal designs are repeated for the whole year using typical weather conditions for Orlando, FL. Figures 3 and 4 show the monthly energy consumption and associated energy cost without ITS and with near optimal and nonoptimal ITS designs. Energy cost is determined by the cost structure introduced before.

By comparing the near optimal ITS design with when there is no ITS installed, it found that monthly energy consumption increases. For example, energy

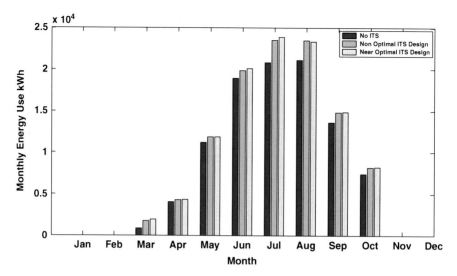

Fig. 3 The monthly energy consumption with/without ice thermal storage ITS

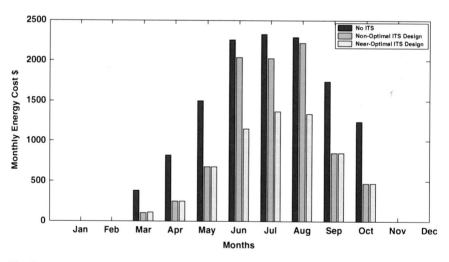

Fig. 4 The monthly energy cost with/without ice thermal storage ITS

consumption rises by using optimal design ITS from 20,822 to 23,899 kWh in July, an increase of 15 %. The main reason for elevated energy consumption with ITS is because of the low chiller efficiency as it operates at lower chilled water-glycol temperature (25 °F vs. 45 °F) (−4 °C vs. 7 °C) to make ice. However, because most of the energy consumption occurred during the off-peak period when the cost of energy is low, the energy cost drops significantly. The energy cost drops from $2329 to $1367 in July, a saving of $962 (41 %). By adding up the monthly energy use and associated cost, the annual cooling energy consumptions with and without

ITS are 108,590 kWh and 97,977 kWh; and the annual cooling energy costs are $6210 and $12,548, respectively. These results indicate that the annual energy consumption increases by 11 % and the energy cost drops by 50 % by using ITS.

By comparing the near optimal ITS design with nonoptimal design, it found that the energy cost drops in most of months and total energy cost drops from $8630 to $6210. The optimal design could provide saving up to 28 % comparing to nonoptimal design. It should be noted that this result is based on one location and a specific cost structure and the readers should not draw a general conclusion on the amount of the operating cost saving indicated in this chapter. The saving varies from one location to other and strongly depends on local utility cost structure. However, the optimization process as shown in Fig. 1 can always be a useful tool to achieve the optimal design as long as the utility cost structure, hourly cooling loads, and outdoor air conditions are correctly entered.

Conclusion

ITS is a promising technology to reduce energy costs by shifting the cooling cost from on-peak to off-peak periods. The ITS can have high impact on energy consumption, demand, and total energy cost. The chapter introduces a tool for central plant optimal design including chiller and ice storage sizes and charging and discharge times. The building energy simulation model eQUEST® is used to generate the hourly cooling load, and the chiller model is used to determine the chiller power. A specific local utility cost structure and one location is used as example for energy cost analysis. The results demonstrated that although the energy consumption increases by using ITS, the energy cost drops significantly, mainly depending on the local utility rate structure. It showed that a significant cost energy saving can be obtained by optimal ice storage design through using the tool proposed in this chapter. The saving could be up to 28 % compared to nonoptimal design of ITS. The results also indicated that annual energy consumption increased by 11 % and the energy cost dropped by 50 % compared to the case when no ITS is installed. This study focuses on a particular cost structure and climate, local utility cost structure needs to be checked in order to determine if the ITS is cost-effective.

References

ASHRAE Standard 90.1. (2004). *Energy standard for buildings except low-rise residential buildings*. ANSI/ASHRAE/IESNA Standard 90.1-2004.
Deb, K. (2001). *Multi-objective optimization using evolutionary algorithms*. New York: Wiley.
Dincer, I. (2002). On thermal energy storage systems and applications in buildings. *Energy and Buildings, 34*, 337–388.
Goldberg, D. E. (1989). *Genetic algorithms in search, optimization, and machine learning*. Reading, MA: Addison-Wesley.

Kusiak, A., Tang, F., & Xu, G. (2011). Multi-objective optimization of HVAC system with an evolutionary computation algorithm. *Energy, 36*(5), 2440–2449.

MacCracken, M. (2003). Thermal energy storage myths. *ASHRAE Journal, 45*(9), 36–42.

MacCracken, M. (2004). Thermal energy storage in sustainable buildings. *ASHRAE Journal, 46*(9), S2–S5.

MATLAB. (2013). *MATLAB R2013a*. MathWorks. Retrieved from www.mathworks.com

Montgomery, R. D. (1998). Ice storage system for school complex. *ASHRAE Journal, 40*(7), 52–56.

Mossolly, M., Ghali, K., & Ghaddar, N. (2009). Optimal control strategy for a multi-zone air conditioning system using a genetic algorithm. *Energy, 34*(1), 58–66.

Nassif, N. (2012). Modeling and optimization of HVAC systems using artificial intelligence approaches. *ASHRAE Transactions, 118*(2), 133.

Nassif, N., Kajl, S., & Sabourin, R. (2005). Optimization of HVAC control system strategy using two-objective genetic algorithm. *HVAC&R Research, 11*(3), 459–486.

Nassif, N., Tesiero, R., & Singh, H. (2013, Winter). Impact of ice thermal storage on cooling energy cost for commercial HVAC systems. In *ASHRAE meeting*, Dallas, TX.

Sebzali, M. J., & Rubini, P. A. (2007). The impact of using chilled water storage systems on the performance of air cooled chillers in Kuwait. *Energy and Buildings, 39*, 975–984.

Silvetti, B. (2002). Application fundamentals of ice-based thermal storage. *ASHRAE Journal, 44*(2), 30–35.

Solberg, P., & Harshaw, J. (2007). Ice storage as part of a LEED® building design. *Trane: Engineers Newsletter, 36–3*, 1.

Xu, X. H., Wang, S. W., Sun, Z. W., & Xiao, F. (2009). A model-based optimal ventilation control strategy of multi-zone VAV air-conditioning systems using genetic algorithm. *Applied Thermal Engineering, 29*(1), 91–104.

Yau, Y. H., & Rismanshi, B. (2012). A review on cool thermal storage technologies and operating strategies. *Renewable and Sustainable Energy Reviews, 16*, 787–797.

Zhou, J., Guanghua, W., Deng, S., Turner, D., Claridge, D., & Contreras, O. (2005). Control optimization for chilled water thermal storage system under complicated time-of-use electricity rate schedule. *ASHRAE Transactions, 111*(1), 184–797.

Optimization of HVAC Systems Using Genetic Algorithm

Tony Nguyen and Nabil Nassif

Abstract Intelligent energy management control system (EMCS) in buildings offers an excellent means of reducing energy consumptions in heating, ventilation, and air-conditioning HVAC systems while maintaining or improving indoor environmental conditions. This can be achieved through the use of computational intelligence and optimization. The chapter thus proposes and evaluates a model-based optimization process for HVAC systems using evolutionary algorithm. The process can be integrated into the EMCS to perform several intelligent functions and achieve optimal whole-system performance. The proposed process addresses the requirements of the latest ASHRAE Standard 62.1. A whole building simulation energy software is used to generate the sub-hourly load. The simulations are performed to test the process and determine the potential energy savings achieved. The testing results demonstrate that total energy consumed by the HVAC system can be reduced by about 26 % when compared to the traditional operating strategies applied.

Introduction

Great efforts have been invested in minimizing the energy costs associated with the operation of HVAC systems. Intelligent Energy Management and Control Systems (EMCS) can provide an effective way of decreasing energy costs in HVAC systems while maintaining indoor environmental conditions (ASHRAE 2010). This EMCS can include several intelligent functions such as optimum set points and operating modes (ASHRAE 2011; Nassif et al. 2005; Wang and Jin 2000; Zheng and Zaheer-Uddin 1996) and fault detection and diagnosis (Seem 2007; Lee et al. 1996). The intelligent EMCS can be achieved through the use of the computational intelligence and optimization (Hani 2009). The intent of this research is to develop an intelligent strategy using technology to drop energy consumption and evaluate a model-based optimization process for HVAC systems. The HVAC optimization problems are

T. Nguyen • N. Nassif (✉)
Department of Civil, Architectural and Environmental Engineering,
North Carolina A&T State University, Greensboro, NC, USA
e-mail: nnassif@ncat.edu

solved using the genetic algorithm (GA) inspired by natural evolution which is successfully applied to a wide range of applications including HVAC systems (Deb 2001; Goldberg 1989; Xu et al. 2009; Mossolly et al. 2009). A whole building simulation energy software eQuest is used to generate the sub-hourly load in order to evaluate the proposed process and determine the potential energy savings achieved.

Methodology

Figure 1 shows a schematic diagram of a typical HVAC system. A typical HVAC system that uses variable air volume (VAV) control is illustrated in Fig. 1. It consists mainly of (a) the return and supply fans, (b) the outdoor, discharge, and recirculation dampers, (c) the air handling unit (AHU) with components such as filter and cooling and heating coils, (d) the pressure-independent VAV terminal boxes, and (e) the local-loop controllers (i.e., C_1, C_2, and C_3). The supply air

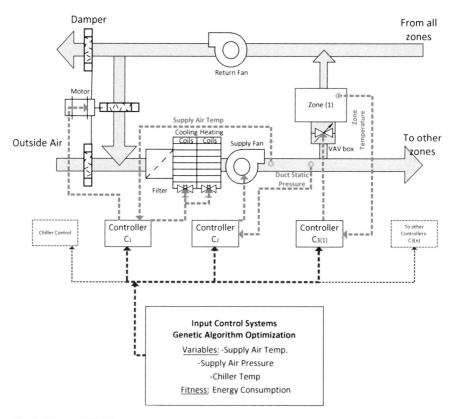

Fig. 1 Typical HVAC system

temperature is controlled by the controller (C_1). The duct static pressure is controlled by the controller (C_2). The zone air temperature at any particular zone n is controlled by the controller ($C_{3\,(n)}$). The proposed process collects the measured data (real data) from components or subsystems. The developed models provide optimal whole-system performance through determining optimal set points and operation sequences. As recommended in this chapter for optimal control strategy, at each time interval (e.g., 15 min), the models provide optimal whole-system performance through determining optimal set points and operation sequences.

A genetic algorithm as shown in Fig. 2 is used to solve the optimization problem. The problem variables are the controller set points and operation modes. In this chapter, the supply air temperature, duct static pressure, and outdoor air flow rate

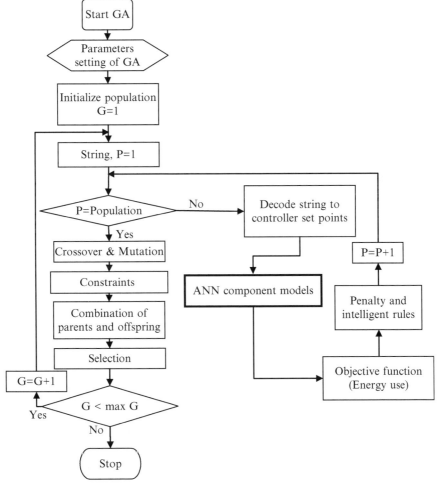

Fig. 2 Flow chart of genetic algorithm GA for the optimization process

are the only problem variables considered. The objective function is the total energy consumption over the optimization period, determined by the component models. The GA starts with a random generation of the initial population (initial solution). The problem variables (controller set points) are encoded to form a chromosome (a string of variables) that represents an individual solution in the population. The performance or objective function f of each individual of the first generation is estimated. The second generation is generated using operations on individuals such as selection, crossover, and mutation, in which individuals with higher performance (fitness) have a greater chance to survive (Deb 2001). The performance of each new individual is again evaluated. The process is repeated until the maximum number of generations (G_{max}) is reached. The simulated binary crossover SBX is used to create two offsprings from two-parent solutions. The random simplest mutation operator is applied to create a new solution from the entire search space. This algorithm uses the elite-preserving operator, which favors elites of a population by giving them an opportunity to be directly carried over to the next generation.

After two offsprings are created using the crossover and mutation operators, they are compared with both of their parents to select two best solutions among the four parent-offspring solutions. The stochastic universal sampling-SUS version of proportionate selection is used (Deb 2001). A penalty is imposed on the objective functions based on the constraints. The constraints are related to the restrictions on the operation of the HVAC system and the rate of variable changes. They cover the lower and upper limits of variables, the design capacity of components. The fan and zone airflow rates, for instance, are restricted within the maximum and minimum limits. The minimum zone air flow rates are restricted to be always higher than 20 % of design air flow rates. The supply air temperature setpoint is limited within 55–65 °F. The duct static pressure setpoint is limited with 1.5 and 2.5 in. Wg. For the results of this study, the computations with the optimization process are performed well within a time length of 3 min on a personal computer. This computation time allows the optimization process to be implemented online. The time could also be decreased using a more modern computer.

Results

The optimization process is evaluated using data generated by whole building simulation energy software for 30,300 ft^2 office building. The cooling and heating is VAV system serving 20 zones. Figure 3 shows the design sensible and total load for those zones. This study focuses on work-time operational hours allowing the zone temperature to drift away during the unoccupied state. Figure 4 shows the building sensible and total load during the occupancy period from 8:00 am (8:00) to 5:00 pm (17:00). As expected, the building load increases as the amount of daylight hours and people increase but decrease near 4:00 pm—when occupants start leaving. Latent load, the difference between total and sensible load, remains relatively constant until 16:00 h when the work hours is approaching closing hours.

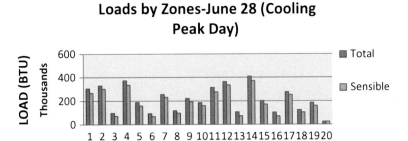

Fig. 3 Loads by zone

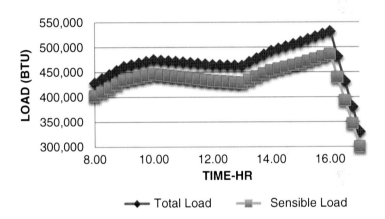

Fig. 4 Loads by time

The optimization process predicts the system performance over a period of 15 min (optimization period). During this short optimization period, the loads and outdoor air conditions are assumed to be constant and estimated from the measured data collected during the previous period. The models are used to find the energy use by each component and then the total energy use in response to the controller set points and operating modes. The inputs are the controller set points (problem variables) and the output is the energy use (objective function). The genetic algorithm GA sends a set of individual solutions containing trial controller set points, and the models then estimate the objective function (total energy use) and send it back to the GA to eliminate, evolve, and pass this solution to the next generation. This process continues until optimal/or near optimal solutions are reached. The supply air temperature, duct static pressure, and outdoor airflow rate are only considered, and the optimization is done only for one summer day as shown in Fig. 4. It is assumed that for the comparison purpose all controller set

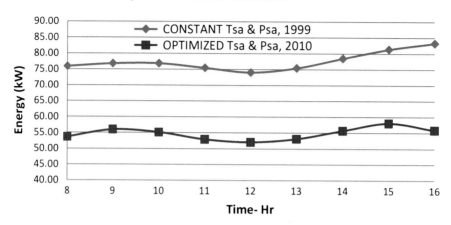

Fig. 5 Optimal and nonoptimal total rate of energy use kW

points are constant for the baseline (nonoptimal condition). In this nonoptimal condition, the supply air temperature setpoint is 55 °F, the duct static pressure setpoint is 2.5 in. Wg, and the ventilation rate is based on ASHRAE Standard 62.1 2001. However, in the optimal condition, the process determines optimally those setpoints and addresses the requirements of the current version of the standard. The optimal supply air temperature setpoint (Tsa) is limited within 55 and 65 °F. The optimal duct static pressure setpoint (Psa) is limited within 1.5 and 2.5 in. Wg. The whole system optimization finds the solution that produces the least total energy use.

Figure 5 shows the optimal and nonoptimal total rate of energy use. The total energy use is equal to the sum of fan, electric reheat, and chiller powers. As a result, by applying the optimization process, the total energy saving is 26 % for this day. The savings could vary depending on the system types, building type and locations, existing energy efficiency opportunity and current control strategies.

Conclusion

A GA algorithm is proposed for the use in HVAC control and operation. The proposed process based on the current version of ASHRAE Standard 62.1. The process can be integrated into the EMCS to perform several intelligent functions and achieve optimal whole-system performance. A whole building simulation energy software is used to generate the sub-hourly load. The simulations are performed to test the process and determine the potential energy savings achieved.

The testing results demonstrate that total energy consumed by the HVAC system can be reduced by about 26 % when compared to the traditional operating strategies applied. The saving could vary depending on the system types, building type and locations, existing energy efficiency opportunity, and current control strategies.

References

ASHRAE. (2010). *ASHRAE standard 62-2010, ventilation for acceptable indoor air quality*. Atlanta, GA: American Society of Heating, Refrigerating and Air-Conditioning Engineers.

ASHRAE. (2011). *ASHRAE handbook-HVAC applications*. Atlanta, GA: American Society of Heating, Refrigerating and Air-Conditioning Engineers. Chapter 42.

Deb, K. (2001). *Multi-objective optimization using evolutionary algorithms*. New York: Wiley.

Goldberg, D. E. (1989). *Genetic algorithms in search, optimization, and machine learning*. Reading, MA: Addison-Wesley.

Hani, H. (2009). Employing computational intelligence to generate more intelligent and energy efficient living spaces. *International Journal of Automation and Computing, 5*(1), 1–9.

Lee, W. Y., House, J. M., Park, C., & Kelly, G. E. (1996). Fault diagnosis of an air handling unit using artificial neural networks. *ASHRAE Transactions, 102*, 540–549.

Mossolly, M., Ghali, K., & Ghaddar, N. (2009). Optimal control strategy for a multi-zone air conditioning system using a genetic algorithm. *Energy, 34*(1), 58–66.

Nassif, N., Kajl, S., & Sabourin, R. (2005). Optimization of HVAC control system strategy using two-objective genetic algorithm. *HVAC&R Research, 11*(3), 459–486.

Seem, J. E. (2007). Using intelligent data analysis to detect abnormal energy consumption in buildings. *Energy and Buildings, 39*, 52–58.

Wang, S., & Jin, X. (2000). Model-based optimal control of VAV air-conditioning system using genetic algorithm. *Building and Environment, 35*, 471–487.

Xu, X. H., Wang, S. W., Sun, Z. W., & Xiao, F. (2009). A model-based optimal ventilation control strategy of multi-zone VAV air-conditioning systems using genetic algorithm. *Applied Thermal Engineering, 29*(1), 91–104.

Zheng, G. R., & Zaheer-Uddin, M. (1996). Optimization of thermal processes in a variable air volume HVAC system. *Energy, 21*(5), 407–420.

Artificial Intelligent Approaches for Modeling and Optimizing HVAC Systems

Raymond Tesiero, Nabil Nassif, and Harmohindar Singh

Abstract Advanced energy management control systems (EMCS) offer an excellent means of reducing energy consumption in heating, ventilating, and air conditioning (HVAC) systems while maintaining and improving indoor environmental conditions. This can be achieved through the use of computational intelligence and optimization. This research will evaluate model-based optimization processes for HVAC systems utilizing MATLAB's neural network Toolbox, which minimizes the error between measured and predicted performance data. The process can be integrated into the EMCS to perform several intelligent functions achieving optimal system performance. The development of a neuron model and optimizing the process will be tested using data collected from an existing HVAC system of a building on the campus of NC A&T State University.

This proposed research focuses on control strategies and Artificial Neural Networks (ANN) within a building automation system (BAS) controller. The controller will achieve the lowest energy consumption while maintaining occupant comfort by performing and prioritizing the appropriate actions. Recent technological advances in computing power, sensors, and databases will influence the cost savings and scalability of the system. Improved energy efficiencies of existing Variable Air Volume (VAV) HVAC systems can be achieved by optimizing the control sequence leading to advanced BAS programming. The program's algorithms analyze multiple variables (humidity, pressure, temperature, CO_2, etc.) simultaneously at key locations throughout the HVAC system (pumps, cooling tower, chiller, fan, etc.) to reach the function's objective, which is the lowest energy consumption while maintaining occupancy comfort.

R. Tesiero, Ph.D.
Department of Computational Science and Engineering, North Carolina A&T State University, Greensboro, NC, USA

N. Nassif, Ph.D., P.E. (✉) • H. Singh, Ph.D., P.E.
Department of Civil and Architectural Engineering, North Carolina A&T State University, Greensboro, NC, USA
e-mail: nnassif@ncat.edu

© Springer International Publishing Switzerland 2016
G.A. Uzochukwu et al. (eds.), *Proceedings of the 2013 National Conference on Advances in Environmental Science and Technology*,
DOI 10.1007/978-3-319-19923-8_22

Nomenclature

ASHRAE American Society of Heating, Refrigeration and Air Conditioning Engineers
BAS Building automation system
EMCS Energy management control systems
HVAC Heating ventilation and air conditioning
MEP Mechanical, electrical, and plumbing

Introduction

The recent global trend shows as fuel costs rise, improving energy efficiency in buildings is a major concern for owners and building managers. Several reasons are behind the push towards a reduction in energy consumption:

- Energy cost
- Government grants
- Utility rebates
- Carbon footprint awareness (Greenhouse gas emissions)
- LEED certification

Today, electricity is generated mainly from nonrenewable energy sources, and over consumption leads to faster depletion of the energy reserves on earth. Electricity is becoming more expensive and generation of electricity from conventional fuels is extremely damaging to the environment because great quantities of carbon dioxide and monoxide, sulphur dioxide, and other hazardous materials are released into the atmosphere. Reducing the consumption of electricity will prolong the existence of the natural energy reserves and limit pollution of the atmosphere while at the same time save money.

A structured approach to energy management can help to identify and implement the best ways to reduce energy costs for a facility. Today, buildings in the United States consume 72 % of electricity produced, and use 55 % of US natural gas. Buildings account for about 48 % of the energy consumed in the United States (costing $350+ billion per year), more than industry and transportation. Of this energy, heating and cooling systems use about 55 % (HVAC, Ventilation, and Hot Water Heating), while lights and appliances use the other 35 % (Architecture 2012), as shown in Figs. 1 and 2.

Projected world marketed energy consumption in the next 20 years is in the 600+ quadrillion BTU range (Mincer 2011), see Fig. 3. Power usage in buildings is often inefficient with regard to the overall building operability. The development of building energy savings methods and models becomes apparently more necessary for a sustainable future.

Artificial neural networks (ANN) is an improved methodology to estimate and accurately control building HVAC systems with minimum energy consumption

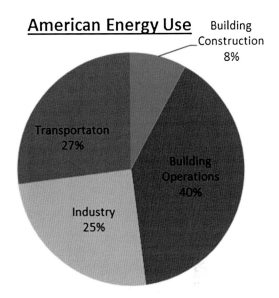

Fig. 1 American energy use

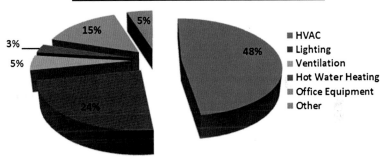

Fig. 2 Total building energy consumption by end use

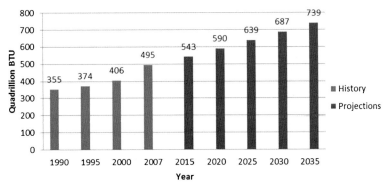

Fig. 3 Projected World marketed energy consumption (www.energy.aol.com)

while maintaining occupant comfort. Additional advantages of ANN methods over other techniques are their faster learning time, simplicity in analysis, and adaptability to changes in a building's energy use. The neural network training process simply involves modification of input variables until the predicted output is in close agreement with the actual output. Previous research utilizing ANNs can achieve 15–20 % savings in building HVAC energy consumption which can equate to 35+ billion dollars and over 100 quadrillion BTUs.

The problems surrounding building energy performance arise from the infinite architectural building designs and multiple energy analysis methods and tools available. Energy efficiency is achieved through properly functioning equipment and control systems, whereas problems associated with building controls and operation are the primary causes of inefficient energy usage. There is an obvious relationship between energy consumption and control-related problems. The most significant problems associated with energy inefficiency are found to be:

- Software
- Hardware
- Equipment maintenance
- Energy management strategies
- Human factors

When a BAS is not present, a more "hands-on" approach is necessary. Training and commitment to control strategies will save money; as long as the building's energy use systems are running properly, the systems can be controlled efficiently. Failure to utilize available features restricts equipment use, especially with controls. Most buildings are using only a small portion of their control capabilities. There are a number of common human factors that contribute to this problem, see Table 1.

For many years, control has been a very active area of the research and development in the HVAC field, aiming at the operation of HVAC systems in terms of reducing overall system operating cost, satisfying thermal comfort of occupants, and ensuring indoor air quality. The increase in energy consumption and demand in the last few decades encourages the investigation of new methods to reduce energy losses. The HVAC systems contribute a significant share of energy consumed in buildings. So it is advisable to find methods to reduce the rise of energy consumption in HVAC system. While there are numerous effective optimal

Table 1 Human factors that waste energy in buildings

Common factors	
Fear of change	Lack of energy conservation awareness from top-down approach
Lack of training	Need to please coworkers' individual comfort levels
Lack of planning	Simplicity of "overriding" system parameters
Insufficient staffing	Lack of fundamental HVAC theory
Fear of internal politics	Lack of programming knowledge
Failure to tune the system	Failure to maintain the system

Table 2 Typical energy management strategies

Strategies	
Time of day scheduling	Chilled/hot water reset
Avoid conservative scheduling	Separate schedules for area or zone usage
Night setback	Zone temperature sensors
Optimal start/stop	Chiller/tower optimization
Implement an energy awareness program	Develop energy competition (NEED)
Economizers	VAV fan pressure optimization
Occupied standby mode	Systems integration
Demand limiting	Demand control ventilation (DCV)
Supply air reset	Variable flow pump pressure optimization

control strategies developed, growing concern for energy efficiency and costs, due to the extremely high fuel oil prices and the shortage of energy supply, has evoked society and building professionals to pay more attention to overall system optimal control and operation and provides incentives to develop the most extensive and robust supervisory and optimal control methodologies for HVAC systems (Wang and Ma 2008).

Regardless of what type of mechanical, electrical, and plumbing (MEP) system exists in a facility, it can be controlled intelligently, effortlessly, and more efficiently with a BAS using typical energy management strategies as shown in Table 2.

These strategies can be implemented without a BAS, using thermostats and/or time control time-of-day schedulers, and a bit of common sense. Typically, a building's single largest expense is energy costs. Utilizing a BAS, to monitor and manage your building's lighting, HVAC, and other systems automatically, and building specific scheduling programs will gain control of energy costs.

Over the last two decades or so, efforts have been undertaken to develop supervisory and optimal control strategies for building HVAC systems thanks to the growing scale of BAS integration and the convenience of collecting large amounts of online operating data by the application of BASs. One of the main achievable goals of the effective use of BASs is to improve the building's energy efficiency, lowering costs, and providing better performance (Wang and Ma 2008).

Energy savings and thermal comfort are important to both facility managers and building occupants. As a result, new innovations in the field are constantly under investigation, including ANN programming. Building performance can be improved with attention to the relationship between design variables and energy performance. Building performance (see Fig. 4) can be divided into three categories:

1. Thermal performance or thermal loads
2. Energy performance or energy-consuming equipment
3. Environmental performance or indoor environmental factors including thermal comfort, lighting, air movement, etc.

Fig. 4 Energy, thermal, and environmental performances of buildings

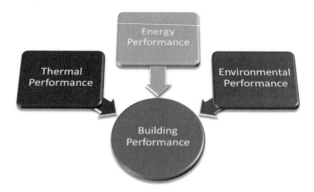

ANNs operate without detailed information about the system. They learn the relationships between input and output variables by studying the historical data. The main advantages of ANNs are their abilities to map nonlinear functions, to learn and generalize by experience, as well as to handle multivariable problems.

During the simulation and optimization calculations, the mathematical model of the HVAC system will include all of the individual component models that influence energy use. Improved energy efficiencies of existing VAV HVAC systems can be achieved by optimizing the control sequence through artificial intelligence using neural networks and genetic algorithms leading to advanced BAS programming. The program's algorithms will analyze multiple variables (humidity, pressure, temperature, CO_2, etc.) simultaneously at key locations throughout the HVAC system (pumps, cooling tower, chiller, fan, etc.) and the program will run simultaneous processes to reach the function's objective which is the lowest energy consumption while maintaining occupancy comfort. The process can be integrated into the EMCS to perform intelligent functions and achieve optimal whole-system performance.

Component models are required for the optimization process. For practical purposes a simple, accurate, and reliable model will be developed to better match the real behavior of the systems over the entire operating range. We will explore the use of an ANN that is data-driven with self-tuning models for HVAC applications, focusing on optimization and control (Nassif 2012a, b). The proposed models and the optimization process will be tested and evaluated using data collected from a typical existing HVAC system in a building on the campus of NC A&T State University.

The component models will be developed and validated against the monitored data. The following required variables will be measured:

- Outdoor and return air temperatures and relative humidity (T_o, T_r, φ_o, and φ_r, respectively)
- Supply air and water temperatures (T_{sa} and T_{sw}, respectively)
- Zone airflow rates ($\dot{Q}_{z,i}$)

- Supply duct, outlet fan, mixing plenum static pressures ($P_{S,sd}$, $P_{S,out}$, and $P_{S,mix}$, respectively)
- Fan speed (N)
- Minimum and principal damper and cooling and heating coil valve positions ($O_{D,min}$, $O_{V,c}$, $O_{D,pr}$, and $O_{V,h}$, respectively)

The following are additional required variables, but they are not measured:

- Fan and outdoor airflow rate ($\dot{Q}_{fan}$ and $\dot{Q}_o$, respectively)
- Inlet and outlet cooling coil relative humidity ($\varphi_{in,c}$ and $\varphi_{ou,c}$, respectively)
- Liquid flow rate through the heating or cooling coils ($\dot{Q}_l$) (Nassif et al. 2004)

Figure 5 shows a schematic diagram of a typical HVAC system and the optimization process integrated into EMCS. Included in this schematic are the proposed ANN models and how they interact within the overall system. It consists mainly of:

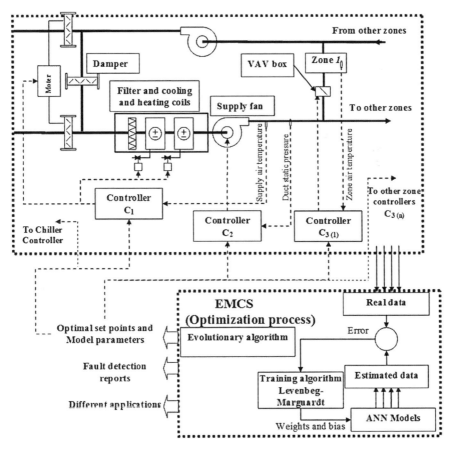

Fig. 5 Typical HVAC system and EMCS along with the proposed ANN models and optimization process (Nassif et al. 2008)

1. The return and supply fans
2. The outdoor, discharge, and recirculation dampers
3. The air handling unit (AHU) with components such as filter and cooling and heating coils
4. The pressure-independent VAV terminal boxes
5. The local-loop controllers (i.e., C_1, C_2, and C_3) (Nassif 2012a, b)

The supply air temperature is controlled by the controller (C_1). The duct static pressure is controlled by the controller (C_2). The zone air temperature at any particular zone n is controlled by the controller ($C_{3\,(n)}$). The EMCS collects the measured data (real data) from components or subsystems. The ANN models are continuously trained using the real data to better match the real behavior of the systems (Nassif et al. 2008). The training algorithm used will be the Levenberg–Marquardt algorithm, which is a built-in algorithm in MATLAB. At each time interval (e.g., 10 min), the ANN models provide optimal whole-system performance by determining optimal set points and operation sequences (Nassif et al. 2008).

In this research, the supply air temperature, duct static pressure, and outdoor air flow rate will be the problem variables considered. The objective function is the total energy consumption over the optimization period, determined by the ANN component models, see Table 3. The HVAC component models based on dynamic artificial neural networks with "self-learning" capability will be developed and utilized in the EMCS to perform the advanced and intelligent functions. The models described here are the fan, cooling coil, and chiller models. Those models can be used for various applications but the inputs and the outputs have to be clearly defined. For the optimization, the model outputs are estimates of the energy consumptions (the objective function) such as fan power and compressor power. The models will be validated and tested for both cases (Nassif 2012a, b).

Table 3 Component models

Component models	Objective function	Independent variables	Dependent variables
VAV model	Energy use and thermal comfort	Outdoor air temperature, indoor sensible loads	Controller set points
Fan model	Fan power	System airflow rate, static pressure	Controller set points
Damper model	Outdoor air flow rate	Outdoor damper position, static pressure	Controller set points
Cooling coil model	Cooling load	Fan airflow rate, entering liquid temperature, entering air DBT, humidity ratio, leaving air DBT	Controller set points
Chiller model	Compressor power	Cooling coil load, chilled water supply temperature, condenser water temperature	Controller set points

Results

Validation of Component Models. The component models that will be developed will then be validated against the monitored variables. The error of the validation models is determined as the absolute ratio of difference between the output and reference variables to the reference variable (Nassif et al. 2004). Data from an existing VAV system will be collected at discrete intervals. The data will be randomly divided into two types of samples: 80 % for training and 20 % for validation. New data will be analyzed for the model testing and to define the model accuracy. The validation samples will be used to measure network generalization and to halt training when generalization stops improving. The testing data will have no effect on training and provide an independent measure of network performance after training. The model performance will be measured by the coefficient of variation CV, which is defined as the ratio of the standard deviation to the mean. The results of the ANN fan, cooling coil, chiller valve, and damper model training, validation, and testing will be graphed and analyzed (Nassif 2012a, b). Preliminary testing results, as shown in Table 4, show how well the models capture the system performance and can be used for the calculations required for the optimization process or any other applications.

The errors of the fan, cooling coil, and chiller models in terms of the coefficient of variation (CV) are within 2–8 %. The models could be incorporated into the EMCS to perform several intelligent functions including energy management and optimal control. MaxE (Maximum Error), MAE (Mean Absolute Error), and CV (Coefficient of Variation).

System Optimization. The optimization process including the dynamic ANN models will be evaluated using data from an existing VAV system. The optimization process will predict the system performance over a period of time. During this short optimization period, the loads and outdoor air conditions are assumed to be constant and estimated from the measured data collected during the previous period. The ANN models will be used to find the energy use by each component

Table 4 Model testing results

Models	Model output (kW)	Model test	Samples	Results		
				MaxE	MAE	CV %
Fan model	Power	Training	34,560 × 3	0.475	0.19	1.13
		Validation	6912 × 3	0.487	0.20	1.2
		Testing	10,080 × 3	0.950	0.227	2.8
Cooling coil model	Cooling load	Training	34,560 × 3	8.56	3.12	4.10
		Validation	6912 × 3	8.72	3.20	4.23
		Testing	10,080 × 3	11.10	5.12	7.8
Chiller model	Power	Training	80	0.81	0.41	2.71
		Validation	20	0.84	0.42	2.12
		Testing	18	1.23	0.58	3.18

and then the total energy use in response to the controller set points and operating modes. The inputs are the controller set points (problem variables) and the output is the energy use (objective function). In the existing system, all controller set points, except the supply air temperature, are constant. The outdoor air is controlled by CO_2 sensor located in a return air duct. The supply air temperature set point is automatically reset based on the supply airflow rate and outdoor air dry bulb temperature. The whole system optimization finds the solution that produces the least total energy use. The total energy use is equal to the sum of fan, electric reheat, and chiller powers (Nassif 2012a, b).

Conclusion

Artificial intelligence approaches are proposed for use in HVAC control and to advance the EMCS. Self-tuning HVAC component models based on an ANN will be developed and validated against data collected from the existing HVAC system. The testing results will show that the models exhibit good accuracy and fit well the input–output data. The errors of the fan, cooling coil, and chiller models in terms of the CV will be calculated. The models could be incorporated into the EMCS to perform several intelligent functions including energy management and optimal control. The testing results will indicate that the optimization process can provide energy saving (Nassif 2012a, b).

References

Architecture, Doerr. (2012). *Sustainability and the impacts of building.* Retrieved 2012, from http://doerr.org/services/sustainability.html

Mincer, S. (2011). *Dealmaking: International electricity woes.* Retrieved 2012, from http://energy.aol.com/2011/08/03/international-electricity-woes/

Nassif, N. (2012a). A robust CO2-based demand-controlled ventilation control strategy for multi-zone HVAC systems. *Energy and Buildings, 45,* 72–81. doi:10.1016/j.enbuild.2011.10.018.

Nassif, N. (2012b). Modeling and optimization of HVAC systems using artificial intelligent approaches. *ASHRAE Transactions, 118*(2), 133.

Nassif, N., Moujaes, S., & Zaheeruddin, M. (2008). Self-tuning dynamic models of HVAC system components. *Energy and Buildings, 40*(9), 1709–1720. doi:10.1016/j.enbuild.2008.02.026.

Nassif, N., Stanislaw, K., & Sabourin, R. (2004). *Modeling and validation of existing VAV system components.* Paper presented at Esim Canadian conference on Building Simulation, June 2004, pp. 135–141.

Wang, S., & Ma, Z. (2008). Supervisory and optimal control of building HVAC systems: A review. *HVAC&R Research, 14*(1), 3–32.

Part V
Waste Management

Experimental Study of MSW Pyrolysis in a Fixed Bed Reactor

Emmanuel Ansah, John Eshun, Lijun Wang, Abolghasem Shahbazi, and Guidgopuram B. Reddy

Abstract Municipal solid waste (MSW) is a potential feedstock for producing transportation fuels because it is readily available using an existing collection/transportation infrastructure and fees are provided by the suppliers or government agencies to treat MSW. The dynamic chemical and physical changes of three organic MSW components of paper, wood, and textile residue during pyrolysis in a fixed bed reactor were characterized. The yields of bio-oil collected at the pyrolysis temperature of 300 °C were 12.0 wt% (wet basis) for textile, 16.3 wt% for paper, and 19.7 wt% for wood. The maximum yields of bio-oil were 52.5 wt% for textile obtained at 700 °C, 57.4 wt% for paper obtained at 600 °C and 64.9 wt% for wood obtained at temperature of 500 °C. The char yields generally decreased with increasing temperature because increased quantities of volatiles from the samples were converted to oil and non-condensable gases. For the pyrolysis temperature from 300 ° to 800 °C, the char yields were between 72.2 and 21.8 wt% (wet basis) for wood, 68.2 and 23.3 wt% for paper and 74.2 % and 22.6 wt% for textile, respectively. The heating values of biochar obtained for all MSW components increased steadily with temperature. The HHVs were from 17.7 MJ/kg at 100 °C to 31.2 MJ/kg at 800 °C for wood, 15.2 MJ/kg at 100 °C to 21.3 MJ/kg at 800 °C for paper and 15.8 MJ/kg at 300 °C to 27.2 MJ/kg at 800°C for textile. Biochar generated from MSW components generally showed an increase in carbon and nitrogen contents with the increase in pyrolysis temperature while oxygen and hydrogen decreased with the temperature.

Introduction

Municipal solid waste (MSW) is commonly called "trash" or "garbage" which includes waste such as tires, furniture, newspapers, plastics, wood waste, textile residues, grass clippings, food, and yard waste (Cheng and Hu 2010).

E. Ansah • J. Eshun • L. Wang (✉) • A. Shahbazi • G.B. Reddy
Department of Natural Resources and Environmental Design, North Carolina Agricultural and Technical State University, 1601 East Market Street, Greensboro, NC 27411, USA
e-mail: lwang@ncat.edu

© Springer International Publishing Switzerland 2016
G.A. Uzochukwu et al. (eds.), *Proceedings of the 2013 National Conference on Advances in Environmental Science and Technology*,
DOI 10.1007/978-3-319-19923-8_23

According to the U.S. Environmental Protection Agency, the total amount of MSW generated in USA each year is currently 250 million tons, which has increased by 65 % since 1980. The amount of MSW generated per capita has increased by more than 20 % since 1980. MSW is considered as a very useful resource of energy. The conversion of MSW to energy can be a competitive solution not only because of negligible costs of feedstock supply but also because of the decrease in the volume of MSW disposed in landfills to minimize the associated environmental problems of gas emissions and leachate production. The 1991 National energy strategy encourages the conversion of MSW to energy. As a result, extensive research has been done on viable mechanisms of generating energy from MSW. One of these mechanisms studied in this research is pyrolysis (Sørum et al. 2001; He et al. 2010; Velghe et al. 2011; Zhou et al. 2013).

Pyrolysis is a thermochemical process in which an organic material such as biomass is heated at a temperature around 400–500 °C in the absence of oxygen to produce char (biochar), noncondensable gases (NCG) (synthesis gas), and condensable vapors or aerosols (van de Velden et al. 2010; Hammond et al. 2011). The condensable vapors are rapidly condensed to form bio-oil which is a mixture of organic chemicals with water. The bio-oil can be further upgraded into marketable liquid fuels (Puy et al. 2011). The yields and properties of each of the three products depend on the operating conditions of the pyrolysis process and the characteristics of the feedstock (He et al. 2010). This research was to investigate the dynamic chemical and physical changes of MSW during pyrolysis to produce bio-oil and biochar in a plug flow reactor.

Experiment

Preparation of MSW samples. Three MSW samples (paper, wood, and textile residue) were obtained from the MSW collected in the Greensboro MSW transfer Station. The paper component in the collected waste consisted of different varieties ranging from news papers, paper towel, cardboard, and label papers. The woody biomass component consisted mainly of wood chips from the hard wood species. The textile component comprised cotton residue from household and lint from commercial laundry. The characterized samples were dried in the sun to remove all moisture content. The waste samples after drying were milled separately in a Thomas Wiley Mill with a 1 mm screen (Thomas Scientific, Swedesboro, NJ) and stored in transparent plastic containers to be used for the experiment.

Pyrolytic reaction unit. An experimental unit as shown in Fig. 1 was set up to investigate the pyrolysis of MSW. Pyrolysis was conducted in a horizontal stainless steel (#316) fixed bed reactor of 300 mm in length and 30 mm in internal diameter. An electric furnace was used to maintain the pyrolysis temperatures. The temperature of the electric furnace was controlled by an inbuilt controller with a K-type

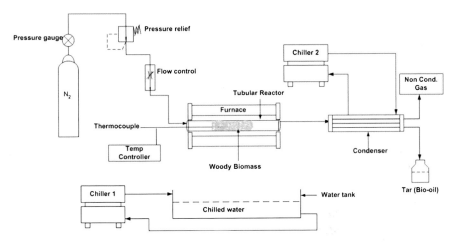

Fig. 1 A simple schematic representation of the fixed bed pyrolysis process

thermocouple. Nitrogen gas was used to purge the air out of the reaction unit. One end of the tubular reactor was connected to the nitrogen gas cylinder by a 1/8 in. (0.3175 cm) stainless steel pipe of 100 mm length. The volumetric flow rate of the purging gas was manually controlled by a rotameter. A K-type thermocouple (1/16 in. sheath) was inserted into the reactor that was filled with the feedstock to measure the actual pyrolysis temperature. The gas outlet of the reactor was connected to three 25 mL vials with stoppers that were connected in sequence in two-stage condensation using cooling water.

Pyrolysis procedure. A total of 5–10 g of MSW components (paper, wood, and textile) were used for each pyrolysis run. The sample was placed in the tubular reactor and it was tightly sealed at both ends using reactor caps. The exact mass of the feedstock was determined by the difference of the mass of the reactor before and after it was filled with the sample. The reactor was heated externally by a thermolyne electric tube furnace placed in a horizontal position at a controlled heat rating. Nitrogen gas at a flow rate of 50 mL/min was used to purge the products out of the reactor. The pyrolysis was conducted in the fixed bed reactor at a temperature from 100 to 800 °C. The temperature of the sample was recorded by a Ni-Cr-Ni thermocouple. Bio-oil and reaction water derived during the pyrolysis were collected in three weighed and labeled 25 mL vials located in the cooling bath. The noncondensable gases were vented through the condenser. The mass of NCG was estimated as the difference from the initial mass of feedstock and the total mass of biochar and condensable bio-oil. After the pyrolysis temperature reached the set value, the reactor was rapidly cooled down to the ambient temperature to stop the reaction by submerging the reactor in a chilled water bath. The biochar sample was then collected and weighed. The experiment for each pyrolysis was repeated three times. The bio-oil and biochar samples were kept in a dark, refrigerated condition at 5 °C. The thermal and physical properties of the biochar and bio-oil were analyzed.

Prior to testing the samples, all bio-oil and biochar samples were removed from the refrigerator and homogenized by vigorously shaking the sample bottle by hand for a minimum of one minute.

Results and Discussion

Effects of pyrolysis temperature on the product distribution. The following data were expressed as the averages of the values that were obtained from replicate measurements. At least three runs were conducted for each experimental condition and at least triplicate measurements were taken for each of the responses. Figures 2 and 3 give the yields of bio-oil and biochar at different pyrolysis temperatures. The yields of volatiles or bio-oil that were condensed and collected at the pyrolysis temperature of 300 °C were 12.0 wt% for textile, 16.3 wt% for paper, and 19.7 wt% for wood on a wet basis. The maximum yields of bio-oil were 52.5 wt% (wb) for textile obtained at 700 °C, 57.4 wt% (wb) for paper obtained at 600 °C, and 64.9 wt% (wet basis) for wood obtained at 500 °C. From the ANOVA analysis, at 95 % confidence interval, the pyrolysis operating temperature within the range of 300–700 °C played a significant role (p-value = 0.002) in the bio-oil production from the MSW components under study.

For the pyrolysis temperature from 300 to 800 °C, the oil yields (on a wet basis) were from 16.3 to 64.9 % for wood, 19.7–57.4 % for paper, and 12–52.8 % for textile, respectively. The yield of bio-oil from the pyrolysis of paper continuously increased with the temperature up to 600 °C and then decreased with further increase of temperature to 800 °C. The yield of bio-oil from the wood pyrolysis increased steadily over the temperature range between 300 and 500 °C, and sharply declined with further increase in temperature up to 600 °C and then increased

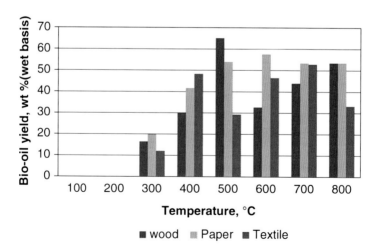

Fig. 2 Effect of temperature on oil yield for three MSW components

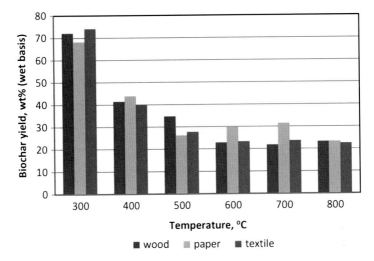

Fig. 3 Effect of temperature on biochar yield for MSW samples

steadily again to 800 °C. The bio-oil yield from the textile pyrolysis increased from 300 to 400 °C and declined from 400 to 500 °C and then increased slightly from 500 to 700 °C with the maximum yield of oil at 700 °C. The results indicated that maximum oil yield from the MSW components were recorded at temperatures between 500 and 700 °C.

Biochar yields for MSW components under study versus pyrolysis temperature are presented in Fig. 3. The char yields generally decreased with increasing temperature because increased quantities of volatiles from the samples were converted to oil and NCG. The char yields for all three MSW components were marked by slight variations over all temperatures. For the pyrolysis temperature from 800 to 300 °C, the char yields were between 21.8 and 72.2 wt% (wet basis) for wood, 23.3 and 68.2 wt% (wet basis) for paper, and 22.6 and 74.2 wt% (wet basis) for textile, respectively.

High Heating Value (HHV) of biochar collected at different temperatures. The heating values of biochar obtained for all MSW components increased steadily with temperature as shown in Fig. 4. The HHVs of biochar from the wood pyrolysis were from 17.7 MJ/kg at 100 °C to 31.2 MJ/kg at 800 °C while the HHVs were 15.2 MJ/kg at 100 °C to 21.3 MJ/kg at 800 °C for paper and 15.8 MJ/kg at 300 °C to 27.2 MJ/kg at 800 °C for textile. It is noted that textile was not pyrolyzed at 100 and 200 °C because of difficulty in collecting the biochar from the tubular reactor at these temperatures. Wood component had higher calorific values at all temperatures than paper and textile, which was consistent with the volatile matter content for the MSW components.

Proximate and ultimate analysis of raw MSW components. Table 1 shows the proximate and ultimate analysis of raw MSW components. The contents of volatiles

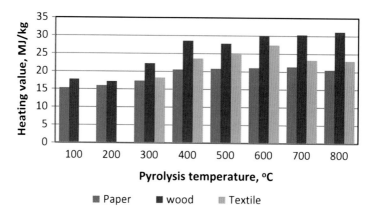

Fig. 4 Heating value of biochar from MSW components from fixed bed pyrolysis at different temperatures

Table 1 Proximate and ultimate analysis of raw MSW components before pyrolysis

Item	Paper	Textile	Wood
Proximate analysis			
Moisture	6.29	4.25	6.57
Volatile	65.62	69.75	73.43
Fixed carbon	21.83	7.12	17.81
Ash	6.26	18.88	2.11
Ultimate analysis			
Carbon	46.0	43.8	45.9
Hydrogen	6.60	6.10	6.67
Nitrogen	1.20	3.5	3.63
Oxygen[a]	45.89	46.2	43.53
Sulfur	0.31	0.30	0.60

[a]Calculated from the difference

of paper, textile, and wood were 65.62 %, 69.75 %, and 73.43 %, respectively. The ash content of textile was as high as 18.88 %. The textile and wood had 3.5 % and 3.63 % of nitrogen while the nitrogen content of paper was only 1.2 %.

Elemental composition of biochar collected at different temperatures. Carbon, hydrogen, and nitrogen were determined using an elemental analyzer operated in the CHN mode. Results obtained for biochar generated from MSW components generally showed an increase in carbon and nitrogen content with the increase in pyrolysis temperature while oxygen and hydrogen decreased with the temperature as shown in Figs. 5, 6, and 7.

As shown in Fig. 5, the carbon content of paper during pyrolysis increased from 41.7 wt% (wb) at 100 °C to 58.8 wt% (wb) at 700 °C, hydrogen decreased from 6.1 wt% (wb) at 100 °C to 0.20 wt% (wb) at 700 °C and oxygen decreased from 53.57 wt% (wb) at 200 °C to 40.03 wt% (wb) at 700 °C. As shown in Fig. 6,

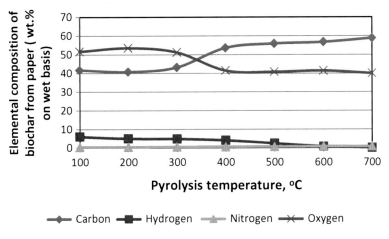

Fig. 5 Elemental composition of biochar fraction of paper pyrolysis

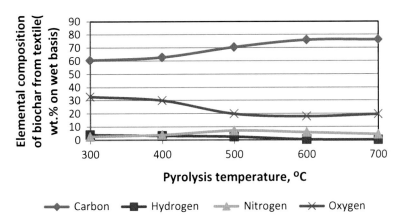

Fig. 6 Elemental composition of biochar fraction from textile pyrolysis

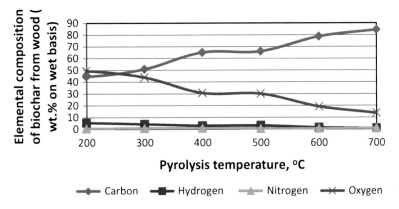

Fig. 7 Elemental composition of biochar fraction from wood pyrolysis

the textile showed relatively high nitrogen content which increased from 2.6 wt% (wb) at 300 °C to 4.3 wt% (wb) at 800 °C. Carbon content increased from 60.5 wt% (wb) at 300 °C to 74.2 wt% (wb) at 800 °C while hydrogen content decreased from 3.9 wt% (wb) at 300 °C to 0.12 wt% (wb) at 800 °C and oxygen content decreased from 32.83 wt% (wb) to 25.1 wt% (wb). As shown in Fig. 7, the carbon content of wood biochar increased from 45.4 wt% (wb) at 100 °C to 84.4 wt% (wb) at 700 °C, hydrogen content decreased from 5.4 wt% (wb) to 0.8 wt% (wb), oxygen decreased from 49.43 wt% (wb) at 200 °C to 13.73 wt% (wb) at 700 °C, and nitrogen increased from 0.4 wt% (wb) to 1.0 wt% (wb).

Conclusion

MSW components (paper, wood, and textile) were pyrolyzed in a 100 mL tubular reactor at different temperatures. Proximate analysis showed that the contents of volatiles of paper, textile, and wood were 65.62 %, 69.75 %, and 73.43 %, respectively. The maximum bio-oil yield was 57 wt% for paper obtained at 600 °C, 52.8 % for textile 700 °C, and 64.9 % for wood obtained at 500 °C. The yield of biochar from the pyrolysis decreased with the increase of temperature. The heating value of biochar obtained from the pyrolysis increased with pyrolysis temperature for all MSW components. The biochar generated from different MSW components generally showed an increase in carbon and nitrogen content with the increase in pyrolysis temperature while oxygen and hydrogen decreased with the temperature. The moisture content of the bio-oil was determined to be very high in the range of 68–72 % at lower pyrolysis temperatures (300 °C) and then decreased with increasing temperature to between 40 and 50 % at 700 °C.

References

Cheng, H., & Hu, Y. (2010). Municipal solid waste (MSW) as a renewable source of energy: Current and future practices in China. *Bioresource Technology, 101*(11), 3816–3824.

Hammond, J., Shackley, S., Sohi, S., & Brownsort, P. (2011). Prospective life cycle carbon abatement for pyrolysis biochar systems in the UK. *Energy Policy, 39*(5), 2646–2655.

He, M., Xiao, B., Liu, S., Hu, Z., Guo, X., Luo, S., et al. (2010). Syngas production from pyrolysis of municipal solid waste (MSW) with dolomite as downstream catalysts. *Journal of Analytical and Applied Pyrolysis, 87*(2), 181–187.

Puy, N., Murillo, R., Navarro, M. V., López, J. M., Rieradevall, J., Fowler, G., et al. (2011). Valorisation of forestry waste by pyrolysis in an auger reactor. *Waste Management, 31*(6), 1339–1349.

Sørum, L., Grønli, M. G., & Hustad, J. E. (2001). Pyrolysis characteristics and kinetics of municipal solid wastes. *Fuel, 80*(9), 1217–1227.

van de Velden, M., Baeyens, J., Brems, A., Janssens, B., & Dewil, R. (2010). Fundamentals, kinetics and endothermicity of the biomass pyrolysis reaction. *Renewable Energy, 35*(1), 232–242.

Velghe, I., Carleer, R., Yperman, J., & Schreurs, S. (2011). Study of the pyrolysis of municipal solid waste for the production of valuable products. *Journal of Analytical and Applied Pyrolysis, 92*(2), 366–375.

Zhou, C., Zhang, Q., Arnold, L., Yang, W., & Blasiak, W. (2013). A study of the pyrolysis behaviors of pelletized recovered municipal solid waste fuels. *Applied Energy, 107*, 173–182.

Separate Hydrolysis and Fermentation of Untreated and Pretreated Alfalfa Cake to Produce Ethanol

Shuangning Xiu, Nana Abayie Boakye-Boaten, and Abolghasem Shahbazi

Abstract The use of lignocellulosic biomass to produce biofuel will add value to land and reduce emissions of greenhouse gases by replacing petroleum products. Valuable co-products derived from fractionation of alfalfa (Medicago sativa) give the resulting fibrous fraction an economic advantage as a feedstock for ethanol production. Freshly harvested alfalfa was dewatered using centrifugation and filtration, whereby alfalfa is separated into a fiber-rich cake and a nutrient-rich juice. Alfalfa solids was pretreated with alkaline soaking (1, 4, and 7 %) at room temperature to evaluate the effects on cellulose digestibility. The production of cellulosic ethanol from alfalfa fibers were investigated by this work using separate hydrolysis and fermentation (SHF). Results show the alkali pretreatment was able to effectively increase cellulosic digestibility of alfalfa solids. A maximal glucose yield of 61 % was obtained with filtered solids with 1 % NaOH pretreatment. The filtration process resulted in a solid fraction with a higher cellulose digestibility, which leads to a higher ethanol production.

Introduction

Declining fossil oil reserves, skyrocketing price, unsecured supplies, and environmental pollution are among the many energy problems we are facing today. These problems necessitate the development of alternative fuels such as biofuels (Xiu et al. 2010). One technology for doing so is the conversion of under-utilized lignocellulosic biomass sources, such as agricultural wastes, forest residues, and dedicated energy crops, into liquid fuel and chemicals that can partially replace petroleum and petrochemicals. However, the recalcitrance of lignocellulosic biomass to chemical and enzyme conversion hinders efficient production of cellulosic

S. Xiu (✉) • N.A. Boakye-Boaten • A. Shahbazi
Department of Natural Resources and Environmental Design, North Carolina A and T State University, 1601 East Market Street, Greensboro, NC 27411, USA
e-mail: xiu@ncat.edu

ethanol. Therefore, a suitable pretreatment process is important to reduce the recalcitrance and to make bioconversion processes more efficient, economic, and environmentally friendly.

Alkaline pretreatment is one of the current leading chemical pretreatment methods, particularly for dissolving lignin. In addition, acetyl groups and various uronic acid substitutes, which lower susceptibility of hemicelluloses and cellulose to hydrolytic enzymes, are also removed by alkaline pretreatment (Mosier et al. 2005). Agricultural residues and herbaceous crops have been shown to be more suitable to alkaline pretreatment than woody biomass (Galbe and Zacchi 2007; Wan et al. 2011). The most commonly used alkali base is NaOH (Li et al. 2004). Alkali pretreatment process has the advantages of utilizing lower temperatures and pressures compared to other lignin removal technologies (Zhang et al. 2010).

Alfalfa is a leguminous perennial crop that does not require either synthetic nitrogen fertilizer or yearly planting and tilling. Thus, the fossil energy inputs to produce a given amount of alfalfa are much less than for annual, non-leguminous species. Alfalfa is widely grown in the USA and has relatively high dry matter yields ranging from about 7 to 23×10^3 kg/ha-year. The technology for growing, harvesting, transporting, and storing alfalfa is already in place in US agriculture. This is an important advantage over any new energy crop.

The primary objective of this ongoing research is to evaluate the efficacy of separation methods and alkali pretreatment on alfalfa solids in terms of enzymatic digestibility. In this study, freshly harvested alfalfa from the North Carolina A&T State University farm was dewatered using centrifugation and filtration. The resulting solid cakes from the two processing methods were collected and processed with or without an alkali pretreatment process, followed by enzymatic hydrolysis. The bacteria Escherichia coli (*E. coli*) was then used to test the fermentability of the sugars enzymatically degraded from alfalfa cellulose.

Materials and Experimental Methods

Grass Harvest and Processing. Two maturities of alfalfa were harvested from existing fields on the NC A&T State University farm. Grass was hand-harvested to an average stubble height of 6 cm for immature and mature alfalfa and immediately transported to the laboratory. Subsamples were taken for assessment of dry matter content. Freshly harvested alfalfa stems were reduced in size using scissors. The biomass was then mixed with water and chopped in a commercial food processor, resulting in a mash with water: biomass ratio of 2:1. Subsequent juice separation was conducted with either a centrifuge (Centra-GP8R Centrifuge, ThermoIEC) or normal filtration (GE WhatmanTM, folded filters, diameter 240 mm). The centrifugation was carried out at a rotational speed of 3600 rpm for 10 min at 25 °C. The resulting juice from the two operations was collected and characterized for its potential in value-added processing and co-products generation. The solid fraction was also dried at 105 °C for 24 h for chemical analysis. Subsamples of the green juice and solid cake were used fresh or kept in a freezer at −80 °C for downstream processing.

To determine how well each separation method would work, the performance of centrifugation and normal filtration was evaluated using the solid content of juices separated as well as the cellulose digestibility. A low solid content of the juice separated and a high cellulose digestibility from the solids fraction are desirable. All the experiments and analyses were performed in duplicate.

Chemical Analyses and Mass Flow Calculation. The alfalfa (parent material) and the solid cake fraction after extracting the juice were analyzed for elemental composition (e.g., C, H, O, N), ash content, solids content, volatile content, and carbohydrates (cellulose, hemicellulose, lignin). Two stages of acid hydrolysis were performed for determining the carbohydrate composition on the alfalfa samples according to NREL Ethanol Project Laboratory Analytical Procedure (Ruiz et al. 1996). The concentrations of ash and carbohydrates in the press juice can be calculated from the proportions of solid fraction and green juice in the alfalfa after separation. In addition to chemical analyses, the mass flow of dry matter from the alfalfa into the green juice and solid cake were calculated. The dry matter of all subsamples of the alfalfa, the solid cake and the green juice was determined by oven-drying at 105 °C for 24 h.

The elemental composition (C, H, O, N) of the alfalfa and the solid cake samples was determined using a PE 2400 II CHNS/O analyzer (Perkin Elmer Japan Co., Ltd.). The solids content analysis was determined using the APHA-AWWA-WPCF Standard Method 2540, which includes total solids (TS), volatile solids (VS), and fixed solids (FS).

Pretreatment of the feedstock. About 50 g of wet alfalfa solid was soaked in 0.25 L of NaOH at various concentrations (1, 4, 7 %) and left at room temperature for 24 h. The mixture was then centrifuged at 2600RCF for 20 min, the supernatant was decanted and the pellet was rinsed with water six times and twice with 0.05 M citric acid buffer (Ph 4.8). Samples were centrifuged and supernatants decanted between rinses.

Enzymatic Hydrolysis. Enzymatic hydrolysis tests were carried out under the same conditions for both the unpretreated and pretreated samples. A control was prepared with an identical amount of raw alfalfa parent material. The total amount of glucose released after 48 h of hydrolysis was measured to calculate the enzymatic digestibility. The conditions of the enzymatic hydrolysis were as follows: about 4.5 g of wet alfalfa was mixed with 0.05 M citrate buffer (pH 4.8) to a total volume of 50 mL. Screw-capped 250-mL Erlenmeyer flasks were used as reaction vessels and were agitated at 180 rpm in a constant temperature rotary shaker at 50 °C for 96 h. The samples were hydrolyzed using a cocktail of enzymes, which included cellulose loading (Novozyme, NS50013) of 25 FPU g/glucan, β-glucosidase (Novozyme, NS50010) at 4.5 CBU/g-glucan, and hemicellulose (Novozyme,

NS22002) at 2.5 FBG/g-glucan. After 96 h, the pretreated slurry was cooled and filtered for sugar determination using HPLC.

Fermentation. The bacterium *Escherichia coli* (*E. coli*) was used to ferment the enzymatically released sugars. For ethanol production, 1 mL of seed culture was used to inoculate 4 mL of Luria-Bertani Broth (LB) medium in a 250-mL Erlenmeyer flask. These were incubated in a shaker at 32 °C and 180 rpm and grown aerobically for 24 h. After 24 h, the cultures were transferred into 50 mL of the medium and the yeast was harvested by centrifugation at 2600 RCF for 15 min and washed with peptone solution three times. The supernatant was discarded, and the cells were transferred into 250-mL Erlenmeyer flasks containing 50 mL of the hydrolysate. The flasks were tightly closed to allow for the fermentation to occur under anaerobic conditions. The cultures were placed in a shaker and incubated at 30 °C for 72 h. Samples were taken at predetermined intervals (0, 3, 24, 72, and 96 h) and collected by filtering through 0.45-µm nylon membranes for ethanol and sugars analysis by HPLC. The ethanol yield was expressed as the percentage of the theoretical yield using the following formula:

$$\text{Yield}_{\text{ethanol}} = \left[\frac{C_{\text{ethanol},f} - C_{\text{ethanol},i}}{0.568 f \cdot C_{\text{biomass}}}\right] \times 100\% \qquad (1)$$

where $C_{\text{ethanol},f}$ is the ethanol concentration at the end of the fermentation (g/L), $C_{\text{ethanol},i}$ is the ethanol concentration at the beginning of the fermentation (g/L), C_{biomass} is the dry biomass concentration at the beginning of the fermentation (g/L), f is the cellulose fraction of the dry biomass (g/g), and 0.568 is the conversion factor from cellulose to ethanol.

Results and Discussion

Mass Flows into Juice and Solid Cake. Approximately 18–27 % of the dry matter contained in the raw alfalfa parent material (PM) was directed into the extracted juice during the centrifuge separation, while 73–82 % was left in the solid cake fraction (Fig. 1). For the filtration process, the mass flow of the dry matter into the juice was between 7 and 16 %, depending on the maturity of the alfalfa. Mature alfalfa results in better separation results with both separation methods than the immature alfalfa.

Compositions of the separated alfalfa solids. The compositions of the separated alfalfa solids are listed in Table 1. The composition of the fresh harvest alfalfa is also reported in Table 1 for comparative purposes. One of the most notable differences between the mature alfalfa and immature alfalfa is the significantly higher total solids content and lignin content of the mature alfalfa. Differences also

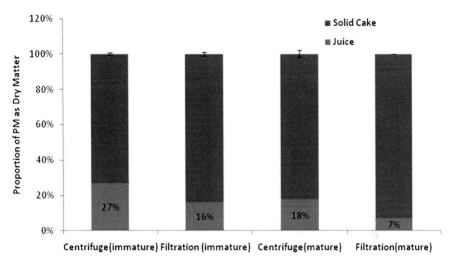

Fig. 1 Mean values of proportions of parent material (PM) that contribute to the mass flow of dry matter (DM) in the juice and solid cake for mature and immature alfalfa with different separation methods

Table 1 Characteristics of raw alfalfa and solid cakes

Group/specific	Mature alfalfa	Immature alfalfa	Immature alfalfa	Immature alfalfa
Separation method	None	None	Centrifuge	Filtration
Solids content				
Total solids, %wt	21.74	16.28	7.29	8.96
Volatile solids, % dry matter	93.01	89.19	91.89	94.28
Ash, % dry matter	6.99	10.81	8.11	5.72
H_2O, %wt	78.26	83.72	82.08	82.30
Composition of dry matter %wt				
Hemicelluloses	17.87	15.59	17.49	6.8
Cellulose	36	35.29	29.3	20.26
Lignin	21.59	14.35	17.78	15.62
Element group (%)				
C	45.06	45.41	45	44.3
H	6.26	6.09	6.39	6.18
O	41.53	42.48	39.9	43.53
N	6.18	5.29	7.76	5.11
S	0.97	0.73	0.95	0.88

existed in the carbohydrates group among these samples. For example, the hemicelluloses content was much lower in immature alfalfa solids with filtration, compared to the other alfalfa, with a value of 6.8 %. The elemental composition is very similar for all of the alfalfa samples.

Enzymatic Hydrolysis

Enzymatic hydrolysis of alfalfa cakes without NaOH pretreatment. Enzymatic hydrolysis was performed to evaluate the cellulose and xylan digestibility of centrifuge solids (CS) and filtered solids (FS) without pretreatment. Raw mature alfalfa was employed as a control. As shown in Fig. 2, the glucose yield increased with enzymatic hydrolysis time and began to level off after 72 h. More glucose was released from separated alfalfa solids than from the control experiment using raw mature alfalfa as a feedstock. The highest glucose yield of 50 % was obtained at 96 h from the FS, 80 % higher than the glucose yield of raw alfalfa. These results suggest that the separation process has a significant impact on the cellulose digestibility of raw alfalfa. The filtration process resulted in a solid fraction with a higher cellulose digestibility.

Figure 3 shows the xylose yield from enzymatic hydrolysis of CS and FS. As seen in Fig. 3, the xylose yield was improved using both separation methods. The FS produced the highest xylose yield, equivalent to 23 % of theoretical yield. Overall, compared to centrifuge separation of alfalfa, the filtration process resulted in higher cellulose and xylose digestibility in the separated solids.

Comparison of alkali pretreatments. The yields of glucose and xylose from enzymatic hydrolysis of alkali pretreated alfalfa solids are shown in Table 2. In comparison with the alfalfa solids without NaOH pretreatment, the glucose yield obtained from both CS and FS with alkaline pretreatment was increased significantly due to a synergistic effect of degradation of hemicelluloses and lignin. However, a slight decrease in glucose yield at higher alkali loading was observed,

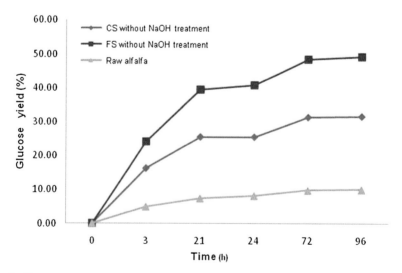

Fig. 2 Glucose yield after enzymatic hydrolysis of centrifuged solids and filtered solids without NaOH treatments (*FS* filtered solid, *CS* centrifuged solid)

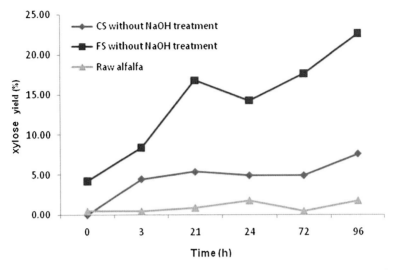

Fig. 3 Xylose yield after enzymatic hydrolysis of centrifuged solids and filtered solids (*FS* filtered solid, *CS* centrifuged solid)

Table 2 Effect of NaOH concentration on the glucose and xylose yields of pretreated alfalfa solids

Biomass type	Sugar yield (%)	0 % NaOH	1 % NaOH	4 % NaOH	7 % NaOH
CS	Glucose	31.71	48.8	49.63	48.05
	Xylose	7.62	21.43	22.75	23.26
FS	Glucose	49.3	60.56	58.59	50.25
	Xylose	22.66	42.88	45.51	46.53

probably due to the high severity pretreatment conditions may lead to undesired sugar loss through dissolution and degradation of hemicelluloses (Chen et al. 2013). The highest glucose of 60.25 % was obtained at 1 % alkali loading for the FS. The xylose yield was increased by alkaline pretreatment as the NaOH concentration was increased. The pretreated FS has higher cellulose digestibility than the pretreated CS samples, which consists with the untreated samples.

Fermentation of Alfalfa Solids for Ethanol Production. The ethanol yield was calculated according to (1). The final ethanol yields for FS and CS without pretreatment were 75 % and 51 %, respectively. These results suggest that the glucose and xylose produced from alfalfa separated solids can be efficiently fermented to ethanol.

Conclusions

1. In comparison with the centrifugal process, normal filtration proved to be more efficient at reducing the solids mass transfer to the juice.
2. The separation process has a significant impact on the cellulose digestibility of raw alfalfa. The filtration process resulted in a solid fraction with a higher cellulose digestibility.
3. Alkali pretreatment improved the enzymatic digestibility of alfalfa solids. Filtered alfalfa solid pretreated with 1 % NaOH produced the highest glucose yield of 61 %.
4. Glucose and xylose released from filtered alfalfa solids can be efficiently fermented to ethanol using *E. coli*, resulting in approximately 75 % of the theoretical ethanol yield.

Acknowledgement The authors are grateful for the support of the USDA-CSREES-Evans-Allen Project, Grant No. NCX-272-5-13-130-1.

References

Chen, Y., Stevens, M. A., Zhu, Y., Holmes, J., & Xu, H. (2013). Understanding of alkaline pretreatment parameters for corn stover enzymatic saccharification. *Biotechnology for Biofuels, 6*, 8.

Galbe, M., & Zacchi, G. (2007). Pretreatment of lignocellulosic materials for efficient bioethanol production. *Biofuels, 108*, 41–65.

Li, Y., Ruan, R., Chen, P. L., Liu, Z., Pan, X., Liu, Y., et al. (2004). Enzymatic hydrolysis of corn stover pretreated by combined dilute alkaline treatment and homogenization. *Transactions of the ASABE, 47*, 821–825.

Mosier, N., Wyman, C., Dale, B., Elander, R., Lee, Y. Y., Holtzapple, M., et al. (2005). Features of promising technologies for pretreatment of lignocellulosic biomass. *Bioresource Technology, 96*, 673–686.

Ruiz, R., & Ehrman, T. (1996). *Determination of carbohydrates in biomass by high performance liquid chromatography* (NREL Chemical Analysis and Testing Standard Procedure, No. LAP-002). Golden, CO: National Renewable Energy Laboratory.

Wan, C., Zhou, Y., & Li, Y. (2011). Liquid hot water and alkaline pretreatment of soybean straw for improving cellulose digestibility. *Bioresource Technology, 102*, 6254–6259.

Xiu, S., Shahbazi, A., Shirley, V. B., Mims, M. R., & Wallace, C. W. (2010). Effectiveness and mechanisms of crude glycerol on the biofuel production from swine manure through hydrothermal pyrolysis. *Journal of Analytical and Applied Pyrolysis, 87*, 194–198.

Zhang, B., Ashahbazi, A., & Wang, L. (2010). Alkali pretreatment and enzymatic hydrolysis of cattails from constructed wetlands. *American Journal of Engineering and Applied Sciences, 3*(2), 328–332.

Tax Policy's Role in Promoting Sustainability

Gwendolyn McFadden and Jean Wells

Abstract While the main objective of a tax system is to raise revenue, when a governmental entity chooses to accomplish an economic or social objective or to encourage a certain behavior or activity, it often uses tax law to accomplish that purpose. Tax law has traditionally and increasingly been used to manipulate the economy. In many ways that is a good thing because in our current green economy, tax law is being used to encourage sustainability and to address other environmental concerns at all governmental levels. Unfortunately, tax professionals may not be optimizing the application of these laws for the benefit of their clients. This chapter provides a brief summary of current and proposed tax incentives that are available to most businesses and individuals. It also describes how public accounting firms are tooling their service options to address the evolving public expectation that businesses will be environmentally responsible world citizens.

Introduction

Although the concept of the environment and accounting are not common bedfellows, the accounting profession is in tune with environmental issues. Historically, Generally Accepted Accounting Principles (GAAP) mandate the reporting and accrual of liabilities such as those that result when man-made disasters such as oil spills and chemical contamination occur. GAAP requirements, pursuant to Statement of Financial Accounting Standards (SFAS) No. 5, Accounting for Contingencies (SFAS5), and SRAS No. 143, Accounting for Asset Retirement Obligations are long standing accounting standards that regulate and mandate this type of reporting. While companies must continue to comply with accounting regulatory standards when liabilities arise, today many companies are becoming socially

G. McFadden (✉)
North Carolina A & T State University, Greensboro, NC, USA
e-mail: mcfaddeg@ncat.edu

J. Wells
Howard University, Washington, DC, USA

responsible citizens and are moving along a course of environmental awareness and volunteerism in their actions and reporting responsibility. Companies and individuals are concerned about conserving energy and protecting the environment. Companies are engaged in establishing strategies leading to environmentally sustainable initiatives. Many are voluntarily disclosing these initiatives to the public in their financial reports which are audited by public accounting firms.

In addition to the practice of voluntarily disclosing environmental sustainable initiatives, companies and individuals are also motivated by tax incentives, particularly tax credits that encourage the purchase and use of energy-efficient products, or the engagement in activities that promote a clean environment. Accounting tax faculty can do their part to promote environmental awareness in the classroom by supplementing classroom discussion with references to tax incentives and tax credits aimed at creating and sustaining a more environmentally friendly world.

Tax Policy and Tax Incentives: In General

While the main objective of a tax system is to raise revenue, when a governmental entity, whether federal, state, or local, chooses to accomplish an economic or social objective or to encourage or discourage certain behavior or activity, it often uses tax law to accomplish that purpose, sometimes in the form of tax incentives. Tax incentives, like direct government expenditures, are subsidies or governmental financial assistance that are grafted on to the income tax revenue generating system. Tax incentives are deeply rooted in the Internal Revenue Code, taking the form of exclusions from income, personal and dependency exemptions, various deductions such as the homeowners mortgage interest deduction, preferential tax rates for capital gains, deferral of tax provisions and, of course, tax credits.

It is well settled that economic and tax policy analysts consider tax incentives to be inefficient. Efficiency, viewed through their lens, is determined by whether or not a tax incentive affects a taxpayer's behavior. The more the tax incentive affects behavior, the more inefficient it is because it is likely to reduce rather than generate revenue. Some public interest advocates have suggested that tax incentives should be removed from the Internal Revenue Code because they involve government outlays in the form of foregone revenues, citing economic inefficiency as a rationale for their position. Others argue that direct tax expenditure is a better means of achieving the same public policy. Notwithstanding the debate, tax incentives promote sound public policy and remain an important, viable and critical component of the Internal Revenue Code.

Sustainability. Sustainability, in its simplest terms, is our capacity for long-term survival and endurance. The United Nations General Assembly adopted a resolution which states that there are three components of sustainability: economic, environmental, and social equity—which are mutually and interdependent pillars (United Nations General Assembly 2007). Environmentalists have long been at the

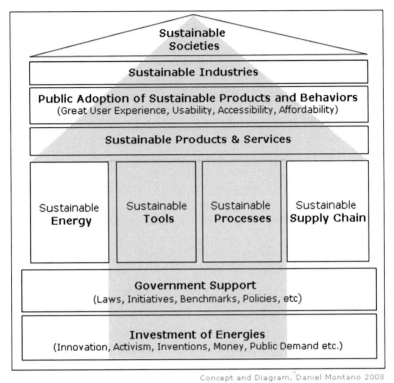

Fig. 1 Sustainable societies diagram. *Source*: Daniel Montano, sustainable societies diagram, 2008

forefront of efforts to develop and promote sustainable projects. In addition, every citizen, manufacturer, industry, product, and service must play a role in ensuring that sustainable societies are maintained as evidenced in the pyramid above in Fig. 1 (Montano 2008). Equally as important is the critical role that government plays by enacting tax laws to incentivize efforts by both individuals and businesses to promote sustainable societies.

Sustainability projects are those that increase the efficiency of machinery and buildings, reduce energy consumption, decrease the negative impact on the environment, and the like. New projects are moving their way to the drawing board. Recently, in June 2013, President Obama announced plans to launch the first federal regulation targeting power plants that emit heat-trapping gases. According to an article in the New York Post, roughly forty percent of carbon dioxide emissions in the United States and one-third of world-wide greenhouse gases can be traced to electronic power plants.

Federal Tax Incentives Related to Sustainability. Roughly 100 years ago, the United States' reliance on petroleum prompted a need to explore and produce oil and gas to support its insatiable supply needs and to reduce its reliance on foreign sources. Tax incentives, in the form of deductions, for depletion and drilling costs were available to companies in the oil and gas business to encourage exploration and production to meet the growing demand. These incentives have stood the test of time and remain a part of today's Internal Revenue Code, but at the same time, the success of these incentives further intensified the country's need and reliance on petroleum. This led to a realization that the United States had to explore alternative sources of fuel and develop ways to conserve energy (Hymel 2006). Former President George W. Bush signed into law the Energy Policy Act of 2005 (2005 Act) which was subsequently amended by the American Recovery and Reinvestment Act of 2009 (2009 Act) to address these concerns.

The 2005 Act included provisions for tax incentives amounting to roughly $14.5 billion over a 10-year period to support the country's infrastructure and address the growing energy needs of the future. Through enhancement of tax incentives already in place, the 2005 Act continued its traditional support of domestic energy production through deduction and allowances for depletion and drilling costs. However, the 2005 Act did much more than that. It signaled Congressional response to the growing concern about the environment. This legislation focused on efforts to promote, enhance, and sustain conservation sources, expand the use of alternative energy sources including wind, solar, ethanol, biomass, hydropower, and encourage development of technology for clean coal.

Intended beneficiaries of the 2005 Act were renewable energy companies, manufacturers of household appliances and homeowners who took advantage of the credits to make energy-efficient improvements to their homes. Indeed, many tax credits that were available to homeowners from 1977 through 1985 for residential alternative energy expenditures had expired in 1985. The 2005 Act law was a gush of wind that propelled Congress' overall efforts to revive and encourage the individual homeowner's efforts to reduce reliance on traditional energy sources.

On February 17, 2009, President Barack Obama signed into law the 2009 Act. This law expanded provisions such as those which allowed a credit for the purchase and installation of energy conservation property such as exterior windows and doors, insulation, heat pumps, furnaces, water heaters, and HVAC units. Similarly, the annual caps for the purchase and use of certain types of property such as the $2000 a year cap for solar equipment, $2000 a year cap for geothermal heat pumps, and $500 a year cap associated with wind turbine usage were eliminated for the tax years 2009 through 2016.

The following table provides a brief summary of the energy credits that are currently included in the Internal Revenue Code. Many were originally enacted during 2005 and have been expanded and renewed by subsequent legislation (Table 1).

States also provide energy tax incentives. Many of these provisions are similar to those available under federal law like those promoting the use of solar equipment and alternative fuel. Others are unique to particular states, such as those encouraging personal recycling.

Tax Policy's Role in Promoting Sustainability

Table 1 Federal energy tax law incentives

Incentive	IRC section	Purpose
For individuals		
Residential energy efficiency improvements credit	Section 25C	Credit for the purchase and installation of qualified energy efficiency improvements and residential energy property
Residential energy-efficient property credit	Section 25D	Credit for the purchase and installation of residential energy-efficient property
For businesses		
Renewable electricity production tax credit	Section 45	Credit for electricity produced from renewable sources
Alternative fuel and alternative fuel mixture credit	Section 40	Credit for using alternative fuel and fuel mixtures
Alternative fuel vehicle refueling credit	Section 30C	Credit for the cost of alternative fuel refueling property
Qualified clean-fuel vehicle property and qualified clean-fuel vehicle refueling property	Section 179A	Deduction for the cost of clean-fuel vehicle and refueling property
Energy-efficient commercial buildings deduction	Section 179	Deduction for the cost of energy-efficient commercial building property
Biodiesel fuels credit	Section 40A	Credit for the production of biodiesel fuels
New energy-efficient home credit	Section 45L	Credit for the construction of energy-efficient homes
Energy-efficient appliance credit	Section 45M	Credit for the manufacture of energy-efficient appliances
Alternative motor vehicle credit	Section 30B	Credit for the production of motor vehicles that increase fuel efficiency
Low sulfur diesel fuel production credit	Section 45H	Credit for the production of low sulfur diesel
Energy credit	Section 48	Credit for energy property placed in service including equipment which uses solar energy to generate electricity, to heat or cool a structure, or to provide solar process heat
Advanced coal project credit	Section 48A	Credit for investment in a qualifying advanced coal project
Gasification project credit	Section 48B	Credit for investment in a qualifying gasification project that uses gasification technology
Advanced energy project credit	Section 48C	Credit for investment in a qualifying advanced energy project

Accounting Firms' Service Options. Today, the scope of services provided by public accounting firms to their clients is much broader than the traditional services or audit, assurance, and tax. Although these services remain the bread and butter of the business, today, major accounting firms are multinational in scope and their services are tailored to meet current and potential demands of their clients.

Although there are many public accounting firms in every locale, the industry is dominated by the Big Four global firms: Deloitte, Ernst & Young (EY), PricewaterhouseCoopers (PwC), and KPMG. A brief overview of the services provided by these firms show that they have all expanded their scope beyond audit, assurance, tax, and consulting services. It is not surprising to find that each, from their unique business perspective, offers services related to environmental sustainability.

Deloitte Sustainability service options include water and other natural resource management and sustainability strategies (Deloitte 2013). EY Climate Change and Sustainability Service line identifies ten service options ranging from the establishment of a sustainability strategy, sustainability reporting advisory and assurance, to greenhouse gas reporting/advisory, and environment due diligence (EY 2013). PwC Sustainable service line focuses on increasing the value of the business, integrating sustainability, and capitalizing on green tax credits (PwC 2013). The KPMG Climate Change and Sustainability Services line focuses on identifying and managing client sustainability risks, capitalizing on tax incentives, and enhancing reporting of corporate social/environmental responsibility and governance (KPMG 2011, 2013).

As you can see, these firms recognize the need to expand the scope of services provided to their clients relating to environmental sustainability initiatives. These firms are finding creative and innovative ways to reach clients on every level and in a rapidly evolving world and marketplace.

Opportunity Missed. Notwithstanding the robust list of service options provided by the Big Four, the flavor is lost if companies are not taking advantage of them. An EY report, issued in 2012, reviewed the connection between the company's tax department and the environmental sustainable initiatives within the company (EY 2013). EY discovered that the connection could be improved.

EY conducted an Environmental Sustainability Tax Survey and received responses from 223 senior executive-level tax directors and Chief Sustainability Officers from various companies (EY 2013). It appears that very few companies have tax departments that are actively involved in establishing environmental sustainability strategies—only 16 % according to the survey. Further, 30 % of the respondents were unaware of the existence of a sustainability leader who could identify tax-incentive opportunities. Only 17 % indicated that their company was aware of and used available incentives related to environmentally sustainable initiatives. EY concluded that businesses were missing out on the opportunity to develop and implement environmentally sustainability strategies. As a result, it is likely that tax incentives to which companies were entitled to would be overlooked. EY's recommendation to address this problem involved improving communication and establishing a more focused approach to identifying tax credit opportunities.

Conclusion

Providing tax incentives as a part of an income tax structure is an inefficient, but good way to encourage and support environmental sustainable initiatives. Many tax incentives, both federal and state, are available to both individuals and businesses. These incentives range from credits for recycling to those that encourage the development of alternative sources of energy. The Big Four public accounting firms have organized their service options to include services that support the environmental sustainable initiatives of their clients, although, based on a survey by one firm, many of those surveyed may be missing the opportunity to take advantage of the tax incentives to which they are entitled.

References

American Recovery and Reinvestment Act of 2009 (Pub. L. 111–5).
Deloitte Sustainability Services. Retrieved August 30, 2013, from http://www2.deloitte.com/us/en/pages/operations/solutions/about-our-sustainability-service.html.
Energy Policy Act of 2005 (Pub. L 109–58).
EY. (2013). Working together: Linking sustainability and tax to reduce costs. Retrieved August 30, 2013 from http://www.ey.com/US/en/Services/Specialty-Services/Climate-Change-and-Sustainability-Services/Working-together---linking-sustainability-and-tax
Hymel, M. (2006). The United States' experience with energy-based Tax incentives: The evidence supporting Tax incentives for renewable energy. *Loyola University Chicago Law Journal, 38*, 43.
KPMG. (2011). Making green greener—tax incentives for energy sustainability. Retrieved August 30, 2013 from http://www.kpmg.com/US/en/IssuesAndInsights/ArticlesPublications/Documents/energy-sustainability.pdf
KPMG. (2013). KPMG green tax index. Retrieved August 30, 2013 from http://www.kpmg.com/global/en/issuesandinsights/articlespublications/green-tax/pages/default.aspx
Montano, D. (2008). Sustainable societies diagram. Retrieved August 30, 2013 from http://multispective.wordpress.com/2008/03/07/sustainable-societies-diagram/
PwC Global Sustainability. Retrieved August 30, 2013 from http://www.pwc.com/gx/en/sustainability/index.jhtml.
United Nations General Assembly. (2005, September 15). 2005 World summit outcome. Resolution A/60/1, adopted by the General Assembly. Retrieved August 30, 2013 from http://data.unaids.org/Topics/UniversalAccess/worldsummitoutcome_resolution_24oct2005_en.pdf

Scrap Tires Air Emissions in North Carolina

Vereda Johnson Williams and Godfrey A. Uzochukwu

Abstract In 2003, more than 290 million scrap tires were generated in the USA. Tires can be used as fuel, either in shredded form known as tire-derived fuel (TDF) or whole, depending on the type of combustion device. Scrap tires are typically used as a supplement to traditional fuels such as coal or wood. Generally, tires need to be reduced in size to fit in most combustion units. Nearly 100 million of these tires were recycled into new products, and 130 million were reused as TDF in various industrial facilities. TDF is one of several viable alternatives to prevent newly generated scrap tires from inappropriate disposal in tire piles, and for reducing or eliminating existing tire stockpiles. Scrap tires represent both a disposal problem and a resource opportunity. Many potential negative environmental and health impacts are associated with scrap tire piles. This study focuses on an examination of North Carolina air emissions related to open tire fires, their potential health impacts, and the combustors that use tires as a fuel.

Introduction

In the USA, the driving force behind control and abatement of scrap tires is stockpiles. Broad state adoption of regulations and regional coordination of neighboring states and local governments have dramatically decreased the incidence of illegal scrap tire storage and disposal. However, local regulations have a limited impact on controlling statewide scrap tire movement and accumulation. Scrap tires can be transported short distances inexpensively, so they usually are moved to the

---

V.J. Williams (✉)
North Carolina A&T State University, Greensboro, NC 27411, USA
e-mail: kingvj@ncat.edu

G.A. Uzochukwu
North Carolina A&T State University, 261 Carver Hall, Greensboro, NC 27411, USA

nearest unregulated jurisdiction or the destination with the lowest disposal cost. Concerns over the costs and hazards associated with large stockpiles as well as the proliferation of new stockpiles have driven most legislation in the USA. *The Scrap Tires: Handbook on Recycling Applications and Management for the U.S. and Mexico* provides a resource for federal, state, and local governments along with private industry in developing markets for the valuable resources contained in scrap tires.

Research Review of Literature on Scrap Tires. A major use for scrap tires is fuel. Tire-derived fuel (TDF) is a fuel derived from scrap tires of all kinds. This may include whole tires or tires processed into uniform, flowable pieces that satisfy the specifications of the end user. Scrap tires are used as fuel, either shredded or whole depending on the type of combustion unit.

TDF is the oldest and most developed market for scrap tires in US industrial facilities across the country, including cement kilns, pulp and paper mills, and electric utilities use TDF as a supplemental fuel to increase boiler efficiency, decrease air emissions, and lower costs. More than 52 % of the 300 million scrap tires generated annually are consumed as TDF in these facilities, providing a cleaner and more economical alternative to traditional fuels.

The U.S. Environmental Protection Agency (EPA) described TDF as a high Btu-value fuel with lower emissions, including lower greenhouse gas emissions, than comparable traditional fuels, in a 2009 Advanced Notice of Proposed Rulemaking. In earlier studies, EPA concluded, "With proper emission controls, burning tires for their fuel energy can be an environmentally sound method of disposing a difficult waste."

Scrap tires make an excellent fuel because of their high heat value. Each tire has energy potential. The heating value of an average size passenger tire is between 13,000 and 15,000 Btu/lb, which compares with about 10,000–12,000 Btu/lb for coal. The primary reason for using tire fuels is to save fuel costs. Further, they are compact, have a consistent composition, and low moisture content—all benefits to the fuel user. Another major reason for combusting tires as fuel is to decrease the number of scrap tires disposed in landfills or stockpiles.

Nationally, scrap tires represent a potential energy source of 1.01 quadrillion Btu per year, based on a discard rate of 300 million tires per year, each weighing an average 22.5 lbs with 15,000 Btu per pound. This is equivalent to 17 million barrels of crude oil and represents about 0.24 % of the US energy needs. Given this energy value, it is clear that scrap tires compete with comparable traditional fuels including coal, petroleum coke, and wood wastes.

While some combustion systems typically cement kilns, can accept tires whole, most combustion systems require the tires to be processed to certain sizes and purity to ensure the material consistently meets the needs of the particular fuel user. Shredding scrap tires to produce TDF uses standard material processing technology which includes shredding and removing dirt or other contaminants.

Processing tires into TDF involves two physical processing steps: chipping/shredding and metal removal. In the first step, tires are either fed into the shredder whole or have the beads removed prior to shredding. The processing equipment is typically high-shear, low-torque shredders. Scrap passenger and truck tires up to 48 in. in outside diameter can be initially reduced in these rotary shear shredders to pieces ranging in size from 1″ to 4″, depending on the end use.

To produce TDF-size shreds and chips, whole tires are reduced to nominal 2-in. pieces using one shredder or a series of shredders, screening equipment and magnetic separation equipment. Magnetic separators are required to remove the steel. A screen in the discharge of the shredder controls the shred/chip size where the 2-in. sized material falls through the screen openings while the oversized material is recirculated back to the shredder. Because a significant amount of rubber is entrained and lost in the wire removal stream, downstream shredding and wire removal can be employed to recover additional rubber, make a cleaner steel product for sale as scrap, and avoid landfilling this wire/rubber material. If smaller-sized TDF (1-in. or crumb rubber) is specified, then more size reduction, metal and fiber separation, classifying, screening, and cleaning equipment may be required.

TDF has the flexibility to be used in a variety of industries. These include cement manufacturing, pulp and paper industry, electric utilities, and tires-to-energy facilities. Cement-manufacturing companies use whole tires and TDF to supplement their primary fuel for firing cement kilns. Several characteristics make scrap tires—either whole or shredded—an excellent fuel for the cement kiln. The very high temperatures and long fuel residence time in the kiln allow complete combustion of the tires. There is no smoke, odor, or visible emissions from the tires. Because the ash is incorporated into the final product, there is no waste. The metal wire contained in the TDF is captured as a raw material or ingredient in the cement-making process. Each passenger car tire contains about 2.5 lbs of high-grade steel. The steel portion of the tire becomes a component of the cement product, replacing some or all of the iron required by the cement-manufacturing process.

The Portland Cement Association (PCA) reports that studies have shown that the use of tires as fuel can reduce certain emissions. According to a 2008 PCA study of emission tests from 31 cement plants firing TDF, there were no statistically significant differences in the emission data sets for sulfur dioxide, nitrogen oxides, total hydrocarbons, carbon monoxide, and metals between kilns combusting TDF and non-TDF firing kilns.

Separate studies conducted by governmental agencies and engineering consulting firms have also indicated that TDF combustion either reduces or does not significantly affect emissions of various contaminants from cement kilns. In a 2007 study, the United States Department of Energy (USDOE) estimated that the combustion of TDF produces less carbon dioxide (CO_2) per unit of energy than coal. This means that when TDF replaces coal in a Portland cement kiln, less CO_2 will be produced.

Long-term experience shows that tires are being used successfully in cement kilns and good quality cement products are being made while using scrap tire fuels. Higher production rates, lower fuel costs, and improved environmental quality achieved when tire fuels are combusted in cement kilns continue to define scrap tires as a viable fuel choice for cement kilns.

Pulp and paper companies use TDF as a supplement to wood waste—the primary fuel used in pulp mill boilers. The technology is proven and has been in continuous use in the USA since the early 1980s. The heating value of the wood waste fuel ranges from about 7900 to 9000 Btu/lb on a dry basis. TDF's higher heat value of 15,000 Btu/lb facilitates uniform boiler combustion, and helps overcome some of the operating problems caused by fuels with low heat content, variable heat content, and high moisture content. The consistent Btu value and low moisture content of TDF and its low cost in comparison to other supplemental fuels make TDF an especially attractive fuel in the pulp and paper industry. In addition, pulp and paper mills have the ability to burn TDF without major equipment modifications—offering yet another advantage to the use of TDF.

Pulp and paper mills continue to increase their use of TDF to help decrease fuel costs and improve both emissions and combustion efficiency. In addition, the use of TDF in pulp and paper mill boilers helps the mills improve their public image in their local regions by demonstrating environmental responsibility. High energy costs, improved reliability in the TDF processing industry, and the consistent product quality of TDF are primary reasons for ongoing growth in both the number of mills consuming TDF and in the amount of TDF consumed per mill.

Electric power utilities use TDF as a supplemental fuel to produce power in boiler operations. Boilers at electric power plants use fuel to generate power for municipalities and industries. In the electric power industry, TDF is used mainly as an additive to other fuels, primarily coal. For electric power utilities, TDF must be correctly sized to fit in fuel conveyors and must be well mixed to ensure proper combustion. Typically, the TDF must be sized at 1-in. × 1-in. and be almost completely dewired for use in the cyclone boilers commonly used in electric power plants. Some electric utilities use stoker-fired units which, because of the longer residence time, can accommodate 2-in. × 2-in. TDF. Smaller, wire-free TDF—50 to 200 mesh—can be used in electric power plants that burn pulverized coal.

In electric power utility applications, TDF provides an economic fuel with constant Btu content and low moisture. Electric utilities also found that the quality of emissions actually improves with the increased use of TDF as a supplemental fuel. Because of its higher heating value, lower emissions, competitive cost, and ability to create stable operating conditions in the boiler, TDF remains an attractive fuel for the electric power generating industry.

For most industrial and institutional boiler systems, TDF sized 2-in. × 2-in. or less and 95 % wire free is an accepted fuel. In industrial boiler applications, combustion of TDF generates energy in the form of steam and/or electricity, displacing the need to generate energy from other power generating facilities and from other fuels, usually coal. This displacement not only offsets the use of certain fuels, it also offsets the pollution emitted from other fuels. TDF combustion in industrial boilers can emit less sulfur dioxide and nitrogen oxide than most types of coal on a net energy output basis. TDF use in industrial boilers remains steady but faces challenges for increased use due to plant closings, depressed markets, and overall economic conditions.

EPA supports the highest and best practical use of scrap tires in accordance with the waste management hierarchy; in order of preference: reduce, reuse, recycle, waste-to-energy, and disposal in an appropriate facility. Disposal of scrap tires in tire piles is not an acceptable management practice because of the risks posed by tire fires, and because of the use of tire piles as a habitat by disease vectors such as mosquitoes. The use of scrap tires as TDF is one of several viable alternatives to prevent newly generated scrap tires from inappropriate disposal in tire piles, and for reducing or eliminating existing tire stockpiles.

EPA testing has shown that TDF has a higher BTU value than coal. Based on over 15 years of experience with more than 80 individual facilities, EPA recognizes that the use of TDF is a viable alternative to the use of fossil fuels, and supports the responsible use of TDF in Portland cement kilns and other industrial facilities provided the candidate facilities have developed a TDF storage and handling plan, have secured a permit for all applicable State and Federal environmental programs, and are in compliance with all requirements of this permit.

The American Society for Testing and Materials (ASTM) approved ASTM 6700-01, an International Standard for TDF, in 2006. ASTM Standard D-6700-01 "Standard Practice for Use of Scrap Tire-Derived Fuel" offers end users and potential end users an industry-accepted standard against which they can compare all tire chips.

Scrap Tire Disposal Problems

Tire piles are excellent breeding grounds for mosquitoes. Because of the shape and impermeability of tires, they may hold water for long periods of time, providing sites for mosquito larvae development. In the southern USA, two exotic species predominate in tires. These two species (*Aedes aegypti* and *Aedes albopictus*) are known to be the principal vectors of Yellow Fever and Dengue disease which afflict millions of people in the tropics. In temperate regions of North America, *Aedes triseriatus* (the native "Eastern Treehole Mosquito") and *Aedes atropalpus* predominate in scrap tires.

Tire stockpiles also have contributed to the introduction of nonnative mosquito species when used tires are transported to the USA. The new species are often more difficult to control and spread more disease. *Aedes albopictus* (the "Asian Tiger Mosquito") merits special consideration. This species was accidentally transported from Japan to the western hemisphere in the mid-1980s in shipments of used tires. It has since become established in at least 23 states. It is considered the nation's most dangerous species.

It is obvious that the elimination of scrap tire piles will eliminate a prolific mosquito habitat along with the associated disease risks. It is also clear that the spread of the Asian Tiger Mosquito has been hastened by interstate shipments of scrap tires. Many states have banned importation of scrap tires for this reason. If elimination of tire piles is not feasible, mosquito abatement programs may be required to suppress mosquito populations at tire piles. This task is problematic and costly, particularly for large piles.

Waste tires and waste tire stockpiles are difficult to ignite. However, once ignited, tires burn very hot and are very difficult to extinguish. This is due to the 75 % void space present in a whole waste tire, which makes it difficult to quench the tires with water or to eliminate the oxygen supply. In addition, the doughnut-shaped tire casings allow air drafts to stoke the fire.

A large tire fire can smolder for several weeks or even months, sometimes with dramatic effect on the surrounding environment. In 1983, a seven-million tire fire in Virginia burned for almost 9 months, polluting nearby water sources.

The air pollutants from fires include dense black smoke which impairs visibility and soils painted surfaces. Toxic gas emissions include polyaromatic hydrocarbons, CO, SO_2, NO_2, and HCl. The heat from tire fires also causes some of the rubber to break down into an oily material. Prolonged burning increases the likelihood of surface and groundwater pollution by the oily material. Using water to extinguish a tire fire is often a futile effort, since an adequate water supply is frequently unavailable. Smothering a tire fire with dirt or sand is perhaps the best current option for extinguishing tire fires. The sand or dirt is moved with heavy equipment to cover the burning tires. This technique does not contribute as greatly to the oil run-off problem and is generally faster and cheaper than foams or water.

According to the EPA in 1992, emissions from open tire fires have been shown to be more toxic than emissions from an incinerator, regardless of the type of fuel. Airborne emissions from open tire fires can have a serious impact on health and the environment. Open tire fire emissions include "criteria" pollutants, such as particulate, carbon monoxide (CO), sulfur oxides (SO_x), nitrogen oxides (NO_x), and volatile organic compounds (VOCs). They also include "non-criteria" hazardous air pollutants, such as polynuclear aromatic hydrocarbons (PAHs), dioxins, furans, hydrogen chloride, benzene, polychlorinated biphenyls (PCBs); and metals, such as arsenic, cadmium, nickel, zinc, mercury, chromium, and vanadium.

Data from a laboratory test program conducted by the EPA in 1993 showed that open tire fire emissions contain 16 times more mutagenic compounds than from residential wood combustion in a fireplace, and 13,000 times more mutagenic compounds than coal-fired utility emissions with good combustion efficiency and add-on controls. The emissions from an open tire fire can pose significant short-term and long-term health hazards to nearby persons (e.g., firefighters and residents). These health effects include irritation of the skin, eyes, and mucous membranes, respiratory effects, central nervous system depression, and cancer.

Scrap Tire Opportunities

Like fossil fuels such as coal, oil, and natural gas, tires contain hydrocarbons. Pound for pound, tires have more fuel value than coal. Hundreds of millions of used tires are generated annually in the USA. By simply disposing of these tires, we miss an important recycling opportunity: the chance to recover their energy and conserve our resources of fossil fuels.

Cement making is an ideal process for recovering this energy. The intense heat of the kiln ensures complete destruction of the tires. There is no smoke or visible emissions from the tires. In fact, the use of tires as fuel can actually reduce certain emissions.

The environmental benefits of utilizing scrap tires as a supplemental fuel in the Portland cement-manufacturing process are multifold. When whole tires are combusted in cement kilns, the steel belting becomes a component of the clinker, replacing some or all of the iron required by the manufacturing process.

In 2008, Portland Cement member companies completed a study on the impact of TDF firing on cement kiln air emissions. The study's data set included emission tests from 31 of the cement plants presently firing TDF. Dioxin-furan emission test results indicated that kilns firing TDF had emissions approximately one-third of those kilns firing conventional fuels; this difference was statistically significant. Emissions of particulate matter (PM) from TDF-firing kilns were 35 % less than the levels reported for kilns firing conventional fuels (not statistically significant due to the low PM emissions reported for essentially all cement plants). Nitrogen oxides, most metals, and sulfur dioxide emissions from TDF-firing kilns also exhibited lower levels than those from conventional fuel kilns. The emission values for carbon monoxide and total hydrocarbons were slightly higher in TDF versus non-TDF firing kilns. However, none of the differences in the emission data sets between TDF versus non-TDF firing kilns for sulfur dioxide, nitrogen oxides, total hydrocarbons, carbon monoxide, and metals were statistically significant.

Separate studies conducted by governmental agencies and engineering consulting firms have also indicated that TDF firing either reduces or does not significantly affect emissions of various contaminants from cement kilns. Under their program for the voluntary reporting of greenhouse gases, the USDOE has estimated that the combustion of TDF produces less carbon dioxide (CO_2) per unit of energy than coal (USDOE 2007a). This means that when TDF replaces coal in a Portland cement kiln—for example, when scrap tires are used to heat the precalciner vessel instead of coal—less CO_2 will be produced. The Mojave Desert Air Quality Management District in California has determined that TDF use is NO_x RACT ("Reasonably Available Control Technology") for Portland cement kilns.

The European Tire and Rubber Manufacturers reported in 2007 that the use of TDF is common in other parts of the world. In Japan, there were 103 million scrap tires generated in 2006 with 16 % being used as a fuel in the cement industry. For that year, Japan recycled 54 % of all scrap tires through heat utilization (JATMA 2007). Of the approximately 3.2 million metric tons of scrap tires handled annually in Europe, 31.6 % are directed to energy recovery systems including Portland cement kilns.

The number of cement plants utilizing scrap tires as a supplemental fuel has risen dramatically over the last 19 years. As of 2006, state and local environmental agencies have approved the use of TDF at 48 plants in 21 states.

Scrap Tires in North Carolina

All North Carolina counties are required to provide a facility for scrap tire collection and to report on their management programs. In FY 2010–2011, North Carolina businesses and individuals disposed of approximately 152,006 t of tires. These tires were managed by county collection facilities and private processing/disposal facilities as follows: 122,206 t managed by counties and shipped to two NC processing firms, 1508 t were managed by counties and shipped to out-of-state processors, 28,292 t of tires were taken directly to processing firms (privately funded cleanups or tire dealers not participating in a county program).

In addition, the two NC processors received 37,918 t of tires from other states. The tire program's success is illustrated by the number of tires properly disposed at permitted facilities. When free disposal was implemented in 1994 for scrap tires generated in the normal course of business in North Carolina, a potential problem emerged: the illegal free disposal of out-of-state tires at county collection sites. Counties should be diligent in screening scrap tires brought for disposal to identify out-of-state tires and other tires not eligible for free disposal. Those that do not are likely spending a portion of their tire tax revenues for disposal of out-of-state tires. The Solid Waste Section assists counties in learning how to avoid fraudulent disposal of out-of-state tires. County efforts to deter disposal of out-of-state tires are a factor when awarding grants to cover cost over-runs.

There are 97 county programs and one regional program [Coastal Regional Solid Waste Management Authority includes Carteret, Craven, and Pamlico Counties] in North Carolina. The programs reported spending a total of $11,787,479.39 for scrap tire management and disposal. Of this total, $11,198,657.28 was for direct disposal costs and $588,822.11 was for other costs, such as labor or equipment costs. Counties with unusually low costs may be stockpiling tires during the year rather than sending them for processing. Some of the variance is probably due to recordkeeping errors or county reporting errors. Some counties manage tires more efficiently than others. For example, counties that allow citizens to dispose tires at multiple recycling facilities or provide curbside pickup incur increased labor costs to recover and load tires into trailers. An analysis indicates that cost of disposal is an average of $103 per ton of scrap tires at recycling/disposal facilities.

County programs annually report the amount of scrap tire tonnages and costs per ton. The costs are reported on the Section website at: County Reports of Tonnages, Costs, Revenue, Cost per Ton. The information on the website was taken from NC Department of Revenue reports of tire tax distribution and from the Scrap Tire Management Annual Reports submitted by the counties.

In FY 2010–2011, 75 % of scrap tires received by North Carolina tire recycling companies were recycled. In order of weight recycled, the categories are: tire-derived fuel, crumb/ground rubber, civil engineering (including drain field material), recap/resale, and other products. The remaining tires go to the two permitted tire monofills in the state. The market for TDF, strong in recent years, saw a 12 % decrease during FY 2010–2011 from the FY 2009–2010 amount.

Table 1 Companies that accept "tires" from NC for burning for tire-derived fuels.Source: 2011–2012 N.C. Department of Revenue Scrap Tire Management Annual Reports

Company name	City
Advanced Disposal Systems of the Carolinas	Charlotte
All Points Waste Service	Garner
Brays Recapping Service	Mount Airy
Carolina Disposal Service, Inc.	Lexington
Central Carolina Tire Processing	Cameron
Elastrix, LLC	Pilot Mountain
Gladden Tire Disposal	Charlotte
Jackson Paper Manufacturing Co.	Sylva
Junk Rescue Company	Charlotte
New East Recycling and Container Service, Inc.	Greenville
PalletOne, Inc. Nationwide Post Industrial Recycling Division	Newton
Polymer Technologies, Inc.	Raeford
Rubber Mulch Unlimited	Mooresville
Trashies Waste Management	Stella
U.S. Tire Recycling, L.P.	Concord
We Recycle Tires	Salisbury
White's Tire Service of Wilson, Inc.	Wilson

Source: 2011-2012 N.C. Department of Revenue Scrap Tire Management Annual Reports

In FY 2011–2012, the total collection of scrap tires in NC was 144,461.8 t. The total revenue from these tires was $11,889,644.95 while the cost of collecting these tires was $14,737,202.67. The average cost per ton for the state of North Carolina was $103.43 per ton. Approximately 10 % of all counties in NC have costs higher than the average cost per ton. The following companies accept "tires" from NC for burning for tire-derived fuels (Table 1).

Recommendations

> The mutagenic emission factor for open tire burning is the greatest of any other combustion emission studied. Studies by found that tire burning is 3–4 orders of magnitude greater than the mutagenic emission factors for the combustion of oil, coal, or wood in utility boilers.

Field sampling data from uncontrolled open tire fires is lacking. This is a result of the inherent difficulties encountered in obtaining the data due to safety concerns and the variable nature of the event (e.g., fire size and duration, meteorological conditions, terrain effects, combustion conditions and fire-fighting activities). Furthermore, the primary concern on the part of officials in charge is to provide for the safety and welfare of those who may be affected by the heat and smoke from the fire.

Whether burning TDF in a new facility or as a modification to an existing facility, several issues must be considered. One consideration is the need to convert scrap tires into a useable fuel. This requires a system to dewire, and shred, or otherwise size the tires so they can be accommodated by a combustor. In addition to aiding in feeding, the sized fuel generally allows for more efficient combustion. However, some large combustor configurations, such as cement kilns, wet-bottom boilers, and stoker-grate boilers can be modified to accept whole tires. Modifications to hardware, combustion practices, and/or other operating practices may also be necessary in order to burn TDF. These modifications are case specific, and must be addressed by engineering staff when considering using TDF.

References

ASTM 6700-01 "Standard Practice for Use of Scrap Tire-Derived Fuel". (2007, July). *End of life tyres management in Europe*, European Tyre and Rubber Manufacturers' Association, Brussels, Belgium. Retrieved from http://www.etrma.org/public/activitiesoflttrenf.asp

Environmental Protection Agency. (2010, December). *The scrap tires: Handbook on recycling applications and management for the U.S. and Mexico*, Environmental Protection Agency, EPA 30-10-010.

The Japan Automobile Tyre Manufacturers Association, Inc. (2007). Retrieved from http://www.jatma.or.jp

Lemieux, P. M., & DeMarini D. (1992, July). Mutagenicity of emissions from the simulated open burning of Scrap Rubber Tires, U.S. Environmental Protection Agency, Control Technology Center, Office of Research and Development, EPA-600R-92-127 (NTIS PB-92-217009).

Index

A

Adaptive extended Kalman filter (AEKF)
 advection-dispersion equation, 77, 78, 80
 BTCS, 76, 77, 82
 EKF, 78, 80–82
 ICS-GC, 79, 82, 84
 mathematical formulation, 75–76
 nonlinear matrices, 78–79
 RMSE, 79–80, 82, 83
 system model error, 77–78
 transport processes, 77
Addai, E.B., 75–84
Alaskan regional climate changes
 CCSM4 simulation (*see* Community Climate System Model 4.0 (CCSM4))
 CMIP5 simulation, 49
Allen, J.R., 14
Al-Nabulsi, A.A., 121, 122, 126
American Society for Testing and Materials (ASTM), 253
Ansah, E., 223–230
Artificial neural networks (ANN)
 advantages, 212, 214, 218
 building performance, 216–218
 component models, 218
 HVAC system and EMCS, 217–218
 optimization process, 219–220
Aruscavage, D., 127
ASHRAE Standard 62.1, 207–208
Assumaning, G.A., 63–72
ASTM. *See* American Society for Testing and Materials (ASTM)
Awaisheh, S.S., 119–128

B

Backward in time and centered in space (BTCS), 76, 77, 82, 83
Barrett, A.P., 26
BAS. *See* Building automation system (BAS)
Baseflow Filter program, 110
Beuchat, L.R., 120, 127, 128
Bhatt, U., 47–59
Bifidobacterium stains
 antibiotic drugs, 172
 arthritis, hypertension, and diabetic medications, 172
 average population, 173–176
 bacterial enumeration, 173
 culture preparation and activation, 173
 drug impact assay, 173
 human health and disease prevention, 171–172
 NSAID drugs, 177
Boakye-Boaten, N.A., 233–240
Boyles, R.P., 14–16
Brackett, R.E., 120
BTCS. *See* Backward in time and centered in space (BTCS)
Buffalo Creek watershed
 ArcGIS tool, 106–108
 Baseflow Filter program, 110
 conductivity values, 113
 finite-difference method, 106
 groundwater depth data, 113–114
 hydraulic heads, 113
 MODFLOW, 108–109, 111–112
 monitoring well, 109
 parameter values, 109

Buffalo Creek watershed (*cont.*)
  PMWIN, 108–110, 112
  site description, 107
  timeframes, 105–106
  treatment plants, 110–111
  trial and error procedure, 110
Buffering capacity (BC), 160, 162–163
Building automation system (BAS), 215, 216

C
Cardiovascular disease (CVD), 180, 181
CBHAR. *See* Chukchi-Beaufort High-resolution Atmospheric Reanalysis (CBHAR)
CCSM4 simulation. *See* Community Climate System Model 4.0 (CCSM4)
Cement kilns, 250, 251, 255
Chang, S.-Y., 63–72, 75–84, 87–94, 97–114
Chan, L.W., 64
Chattopadhyay, S., 3–10
Chen, H., 88
Chen, J., 21–33
Chiang, W.-H., 108
Chilled water plants
  chiller power, 191
  data analysis, 192
  fan power, 192
  methodology, 189
  optimization
    energy simulation, 188–189
    MATLAB program, 188
    vs. nonoptimal conditions, 190
    vs. thumb, 189
  pump power, 192
  supply temperature, 191
  temperature difference, 190
  total power, 191, 192
Chukchi-Beaufort High-resolution Atmospheric Reanalysis (CBHAR)
  climatological circulation, 24–26
  climatology analysis
    intense storms *vs.* strong winds frequency, 29–30
    ninety-fifth-percentile wind speed, 28
    shaping strong wind field, 30–32
    SLP, 31–32
  offshore energy development, 21
  storm climatology and variability, 26–28
  storm identification algorithm
    SLP, 23–24
    WRF model, 22–23

Climate change impact assessment
  direct method
    evapotranspiration, 7–8
    hydrologic simulation, 4, 8–10
    monthly variations, 9
  frequency perturbation method, 5
  GCM outputs, 4
  hydrological response, 6–8
  IPCC report, 3
  Raccoon River watershed
    daily temperature, 6, 7
    monthly analysis, 6
    study area, 4–5
  SWAT, 4
Climate variability, North Carolina
  study area and data, 13–14
  trend analysis
    Mann–Kendall test, 14, 15
    precipitation and temperature, 13
    pre-whitening, 15
    SQMK test, 14–16
    TSA approach, 14–16
Cluster-tree topology, 40
Community Climate System Model 4.0 (CCSM4)
  CORDEX, 48–49
  downscaling domain, 49, 50
  GCM output, 48
  NCEP/NCAR, 48
  RCP6 scenario, 50, 56–58
  twentieth century simulations
    annual mean biases, 53, 54
    Cressman interpolation technique, 53
    monthly mean biases, 53, 54
    mountain lifting effects, 52
    seasonal variability, 51, 52
    surface winds, 51, 53
    temperature, 54–56
    wind speed and precipitation, 54–55
  WRF, 50–51
Continuous pollutant source
  advection-dispersion equation, 99
  background-error covariance, 98–99
  background state vector, 100
  boundary conditions, 99
  contaminant transport, 101–102
  discrete Kalman filter process, 100
  interpolation confidence, 98
  MAE, 103–105
  morpho-dynamic model, 99
  observation equation, 100
  parameters, 101

Index

quasigeostrophic model, 98
RMSE, 102, 103
state-space estimation technique, 98
vector-matrix form, 99
Coordinated Regional Climate Downscaling Experiment (CORDEX), 48–49
Coupled Model Intercomparison Project 5th phase (CMIP5) simulation, 49
Cressman interpolation technique, 53
CVD. *See* Cardiovascular disease (CVD)

D
Decoupled Extended Kalman filter (DEKF), 64
Degaetano, T.A., 14
Domenico, P.A., 89
Donkor, O.N., 168

E
Edmonson, W.W., 35–45
EKF. *See* Extended Kalman filter (EKF)
Energy management control systems (EMCS), 203, 216–218
Energy Policy Act, 2005, 243–244
Ensemble Square Root Kalman filter (EnSRKF)
  advection-dispersion-reaction equation, 88
  analysis perturbation, 91
  EKF, 87–88
  FTCS, 88–89, 92
  Monte Carlo sampling, 90
  observation/measurement equation, 89
  parameters, 91–92
  residual matrix, 90
  RMSE profiles, 92–94
  sequential operation, 89
  state matrix, 90
  system equation, 89
Environmental Protection Agency (EPA), 250, 253
Enzymatic hydrolysis, alfalfa cake ethanol production
  alkaline pretreatment, 238–239
  cellulose digestibility, 238
  fermentation, 239
  xylose digestibility, 238, 239
eQUEST model, 199, 203–204
Escherichia (E.) coli O157:H7
  ascorbic acid and copper
    antimicrobial activity, 132, 134
    bacterial enumeration, 133
    bacterial strains and culture preparation, 132
  sample inoculation, 132
  statistical analysis, 133
  treatment procedure, 133
  treatment solutions, 133–134
  washing fresh produce, 131, 134
Eucalyptus and Wild-thyme, EOs
  antibacterial activity, 122–124, 126
  disinfection procedures, 120, 126
  efficacy, 123
  fresh produce, 119–120, 127
  incidence, 120
  medicinal plants and extraction, 121–122
  MIC values, 125
  pathogen strains and inoculum preparation, 121
  rapid growth rate, 127
  rocket salad leaves, 122, 125–126
  sanitizers, 128
  statistical analysis, 123
Eshun, J., 223–230
Essential Oils (EOs). *See Escherichia (E.) coli* O157:H7
Evapotranspiration (ET), 8, 109, 110
Extended Kalman filter (EKF), 64, 67, 82, 83, 87–88

F
Feldkamp, L.A., 67
Feyyisa, J., 105–114
Fish protein isolates (FPI), 180, 181
Foley, D., 134
Forward-Time and Central-Space (FTCS) method
  continuous pollutant source, 101–102
  EnSRKF, 88, 92
  KF embedded with NN, 65
Freeland, D., 187–201
Frequency perturbation method, 4–5, 8, 10

G
Geer, F.C.V., 98
Generally Accepted Accounting Principles (GAAP), 241–242
Genetic algorithm (GA)
  HVAC system
    ASHRAE Standard 62.1, 5–6
    component models, 206
    computational intelligence, 206
    eQuest model, 203–204
    loads by time, 206, 207

Genetic algorithm (GA) (cont.)
    loads by zone, 206, 207
    optimal and nonoptimal, energy
        use, 208
    optimization period, 207
    optimization process, 205–206
    SBX, 206
    ITS
        constrained/unconstrained optimization problems, 196
        MATLAB®, 198
Ghoshal, T., 87–94
Global climate models (GCMs), 4–6, 47, 51
Ground monitoring wireless sensor network (GM-WSN), 37–38, 41, 42
Groundwater contaminant transport model. See Adaptive extended Kalman filter (AEKF)
Groundwater flow system. See Buffalo Creek watershed
Gums
    bacterial enumeration, 168
    culture maintenance, 167
    media and treatments, 167
    pH values, 167–169
    prebiotics, 166
    probiotics, 166–167
Gyawali, R., 131–134, 171–177

H
Hall, C.L., 187–201
Hamill, T.M., 98
Han, J., 76
Hayek, S.A., 137–142, 145–154, 157–163, 171–177
Heating ventilation and air conditioning (HVAC) system, 204–205
    artificial intelligent approaches
        American energy use, 212, 213
        ANN, 212, 214, 216–218
        BAS, 215
        building performance, 215–216
        component models, 216–218
        electricity, 212
        EMCS, 216–218
        energy efficiency, 212, 214
        energy management strategies, 215
        human factors, 214
        optimal control strategies, 214–215
        power usage, 212, 213
        preliminary testing, 219
        system optimization, 219–220

    EMCS, 203, 208
    genetic algorithm
        ASHRAE Standard 62.1, 207–208
        component models, 206
        computational intelligence, 206
        eQuest model, 203–204
        loads by time, 206, 207
        loads by zone, 206, 207
        optimal and nonoptimal, energy use, 208
        optimization period, 207
        optimization process, 205–206
        SBX, 206
Hendricks, F.H.J., 64
Huang, C., 88
Hu, B., 88
HVAC system. See Heating ventilation and air conditioning (HVAC) system
Hydrologic Response Units (HRUs), 5

I
Ibrahim, S.A., 131–134, 137–142, 145–154, 157–163, 165–169, 171–177, 179–182
Ice thermal storage (ITS)
    applications, 196
    definition, 196
    DOE-2 chiller model, 196
    optimal design
        annual energy cost, 197–201
        central plant piping configuration, 198, 199
        chiller, 197
        energy consumption, 199–201
        eQUEST model, 199
        GA, 196, 198
        HVAC system, 196, 198
        operating strategies, 198
        optimization process, 197, 201
Innovation Covariance Scaling and Gain Correction (ICS-GC), 76, 79, 82–83
Intergovernmental Panel for Climate Change (IPCC), 3–4
Internal Revenue Code, 242
Ipomoea batatas. See *Lactobacillus*
Isoelectric solubilization/precipitation (ISP)
    by-products, 179–180
    functional food product, 180–182
ITS. See Ice thermal storage (ITS)

J
Jha, M.K., 3–10, 13–18, 35–45, 105–114
Jin, A., 64, 65, 76

K
Kalman filter (KF) embedded with neural network (NN)
  distinct phases, 64
  EKF, 67, 69
  FTCS, 65
  initial error covariance, 68
  measurement sensitivity matrix, 67
  model description, 64–65, 68–69
  numerical methods, 63–64
  parameter estimation, 66–67
  prediction techniques
    MAE, 68, 70–71
    RMSE, 68, 70–71
    simulated true value, 69–70
    stochastic Markov chain process, 72
  process and observation models, 66
  recursive and dynamic equations, 67–68
  sparse observation data points, 65
Karlton-Senaye, B.D., 165–169
Karl, T.R., 14
Kinzelbach, W., 64, 108
Knight, R.W., 14
Krieger, J., 47–59

L
Lactic acid (LA), 132, 162
Lactobacillus
  acid phosphatase, 140–142
  α-glucosidase, 138, 140–141
  bacterial enumeration, 138–140
  β-glucosidase, 138, 140–141
  cultivation
    bacterial enumeration, 159, 161–162
    bacterial growth, 160
    culture conditions, 159–160
    fastidious nutritional requirement, 158
    greenish cloud formation, 161
    microorganisms, 158–159
    oleic acid, 158
    pH value and buffering capacity, 160, 162–163
    preparation, 159
    statistical analysis, 160
  culture conditions, 139
  enzyme samples preparation, 139
  fermentation and bioconversion, 137–138
  human nutrition, 138
  L. reuteri (*see Lactobacillus reuteri*)
  media preparation, 138
  phytase, 140–142
  statistical analysis, 140
Lactobacillus reuteri
  acid phosphatase, 148, 150, 152
  α-glucosidase, 147–150
  β-glucosidase, 147–148, 150, 151
  culture conditions, 147
  enzyme samples preparation, 147
  indigestible fibers, 145–146
  media and culture preparation, 147
  nutritional requirements, 146
  phytase, 148, 150, 153, 154
  statistical analysis, 148
Latif, S.M.I., 76, 88, 90, 98
Leunga, C.S., 64
Lignocellulosic biomass, 233–234
Lin, L., 76
Liong, M.-T., 168
Li, S., 64
Lu, C., 47–59

M
MAE. *See* Mean absolute error (MAE)
MATLAB program, 188, 198
McFadden, G., 241–247
McMahon, T.A., 110
Mean absolute error (MAE), 68, 70–71, 102–104
Municipal solid waste (MSW) pyrolysis
  fixed bed reactor
    biochar samples, 225–227
    bio-oil samples, 225–227
    characterized samples, 224
    elemental composition, 228–230
    high heating value, 227, 228
    proximate and ultimate analysis, 227–228
    pyrolytic reaction unit, 224–225
  National energy strategy, 224
  thermochemical process, 224
  trash/garbage, 223

N
Nassif, N., 187–201, 203–209, 211–220
Nathan, R.J., 110
Nguyen, T., 203–209
Nonsteroidal anti-inflammatory drug (NSAID), 177
Nygard, K., 126

O
Obanla, T.O., 171–177

P
Palacios, M.C., 154
Park, C.M., 127
Park, E.E., 76
Parr, A., 80, 91, 98
Portland Cement Association (PCA), 251, 255
Prebiotics, 166
Pre-whitening technique, 15
Probiotics
   Bifidobacterium, 172
   gums
      bacterial enumeration, 168
      culture maintenance, 167
      health benefits, 166–167
      media and treatments, 167
      pH values, 167–169
      prebiotics, 166
   Lactobacillus, 146
Puskorius, G.V., 67

R
Raccoon River watershed (RRW)
   daily temperature, 6, 7
   monthly analysis, 6
   study area, 4–5
Raman, S., 14–16
Rechargeable power supply (RPS-1), 42
Reddy, G.B., 223–230
Representative Concentration Pathway (RCP6), 50, 56–58
Root mean square error (RMSE)
   AEKF, 79–80, 82–83
   continuous pollutant source, 102, 103
   EnSRKF, 92–94
RRW. *See* Raccoon River watershed (RRW)
Russell, D.J., 126

S
Saha, A., 97–104
Satellite Active Passive Mission (SMAP), 36, 38
Sayemuzzaman, M., 13–18
SBX. *See* Simulated binary crossover(SBX)
Schwartz, F.W., 89, 99
Scrap tires air emissions, North Carolina
   county collection facilities, 256
   disposal problems, 253–254
   recycling/disposal facilities, 256
   stockpiles, 249–250
   TDF, 257
      ASTM, 253
      benefits, 250
      cement-manufacturing process, 251
      chipping/shredding, 250–251
      combustion unit, 250
      electric power utilities, 252
      EPA, 250, 253
      industrial and institutional boiler systems, 252
      metal removal, 251
      modifications, 257–258
      PCA, 251, 255
      USDOE, 255
      wood waste fuel, 252
Sea level pressure (SLP), 23–24, 31–32
Selim, S., 127
Sen, P.K., 15
Sen, Z., 67
Separate hydrolysis and fermentation (SHF), alfalfa
   advantage, 234
   E. coli, 234
   ethanol production
      acid hydrolysis, 235
      alkaline pretreatment, 234, 238–239
      cellulose digestibility, 238
      enzymatic hydrolysis, 235–236
      fermentation, 236, 239
      grass harvest and processing, 234–235
      mass flow calculation, 235–237
      PE 2400 II CHNS/O analyzer, 235
      pretreatment, 235
      separated alfalfa solids, 236–237
      xylose digestibility, 238, 239
   lignocellulosic biomass, 233–234
Sequential Mann–Kendall (SQMK) test, 14–15
Serreze, M.C., 26
Shahbazi, A., 157–163, 223–230, 233–240
Simulated binary crossover (SBX), 206
Singh, H., 211–220
SMAP. *See* Satellite Active Passive Mission (SMAP)
Snyder, C., 98, 100
Soil and Water Assessment Tool (SWAT), 4, 5

Index

Soil moisture
  base station, 44
  data flow, 42–44
  design issue, 38
  GM-WSN, 37–38
  ground segment, 38
  network topology, 39–40
  power supply, 42, 43
  RF module, 40–41
  sensor communication, 41–42
  small satellite data retrieval, 36–37
  SMAP mission, 36
  system architecture, 38, 39
  temperature, 36
  5TM sensor, 41, 42
Soil Moisture and Ocean Salinity (SMOS), 36, 38
Special Report Emission Scenario (SRES), 5, 6
Stockpiles, 249–250
SWAT. *See* Soil and Water Assessment Tool (SWAT)
Sweet potato-based medium (SPM). *See Lactobacillus*

T
Tahergorabi, R., 179–182
Tao, W., 21–33
Tax incentive
  Big Four public accounting firms, 245–246
  Deloitte Sustainability service options, 246
  energy credits, 244–245
  Energy Policy Act, 2005, 243–244
  Environmental Sustainability Tax Survey, 246
  Internal Revenue Code, 242
  sustainability, 242–243
Tax policy
  energy-efficient products, 242
  GAAP, 241–242
  incentives
    Big Four public accounting firms, 245–246
    Deloitte Sustainability service options, 246
    energy credits, 244–245
    Energy Policy Act, 2005, 243–244
    Environmental Sustainability Tax Survey, 246
    Internal Revenue Code, 242
    sustainability, 242–243
TDF. *See* Tire-derived fuel (TDF)
Tesiero, R., 211–220

Theil, H., 15
Theil–Sen approach (TSA) test, 14
Thermochemical process, 224
Thornhill, G.D., 99
Three-dimensional (3-D) subsurface contaminant transport model
  EnSRKF (*see* Ensemble Square Root Kalman Filter (EnSRKF))
  KF embedded with NN (*see* Kalman filter (KF) embedded with Neural Network (NN))
  variational analysis
    advection-dispersion equation, 99
    background-error covariance, 98–99
    background state vector, 100
    boundary conditions, 99
    contaminant transport, 101–102
    discrete Kalman filter process, 100
    interpolation confidence, 98
    MAE, 103–105
    morpho-dynamic model, 99
    observation equation, 100
    parameters, 101
    quasigeostrophic model, 98
    RMSE, 102, 103
    state-space estimation technique, 98
    vector-matrix form, 99
Tire-derived fuel (TDF), 257
  ASTM, 253
  benefits, 250
  cement-manufacturing process, 251
  chipping/shredding, 250–251
  combustion unit, 250
  electric power utilities, 252
  EPA, 250, 253
  industrial and institutional boiler systems, 252
  metal removal, 251
  modifications, 257–258
  PCA, 251, 255
  USDOE, 255
  wood waste fuel, 252

U
United States Department of Energy (USDOE), 255
Uzochukwu, G.A., 249–258

W
Walker, D.M., 67
Wang, L., 223–230

Weather Research and Forecasting (WRF) model
    CBHAR, 22
    CCSM4 simulations, 53
    downscaling simulations, 50–51
    RCP6 scenario, 56–58
Wells, J., 241–247
Williams, V.J., 249–258
WRF model. *See* Weather Research and Forecasting (WRF) model

X
XBEE ZB S2 module, 40–41
Xiu, S., 233–240

Y
Yeo, S.-K., 168

Z
Zaman, R., 35–45
Zhang, H., 99
Zhang, J., 21–33, 47–59
Zhang, X., 21–33, 47–59
Zhan, H.H., 76
Zou, S., 80, 91, 98